114 Springer Series in Solid-State Sciences
Edited by Peter Fulde

Springer Series in Solid-State Sciences

Editors: M. Cardona P. Fulde K. von Klitzing H.-J. Queisser

Managing Editor: H. K.V. Lotsch Volumes 1–89 are listed at the end of the book

90 **Earlier and Recent Aspects of Superconductivity** Editors: J. G. Bednorz and K. A. Müller

91 **Electronic Properties of Conjugated Polymers III** Basic Models and Applications
Editors: H. Kuzmany, M. Mehring, and S. Roth

92 **Physics and Engineering Applications of Magnetism** Editors: Y. Ishikawa and N. Miura

93 **Quasicrystals** Editors: T. Fujiwara and T. Ogawa

94 **Electronic Conduction in Oxides** By N. Tsuda, K. Nasu, A. Yanase, and K. Siratori

95 **Electronic Materials** A New Era in Materials Science
Editors: J. R. Chelikowsky and A. Franciosi

96 **Electron Liquids** By A. Isihara

97 **Localization and Confinement of Electrons in Semiconductors**
Editors: F. Kuchar, H. Heinrich, and G. Bauer

98 **Magnetism and the Electronic Structure of Crystals**
By V. A. Gubanov, A. I. Liechtenstein, and A.V. Postnikov

99 **Electronic Properties of High-T_c Superconductors and Related Compounds**
Editors: H. Kuzmany, M. Mehring, and J. Fink

100 **Electron Correlations in Molecules and Solids** 2nd. edn., ed. by P. Fulde

101 **High Magnetic Fields in Semiconductor Physics III**
Quantum Hall Effect, Transport and Optics By G. Landwehr

102 **Conjugated Conducting Polymers** Editor: H. Kiess

103 **Molecular Dynamics Simulations** Editor: F. Yonezawa

104 **Products of Random Matrices** in Statistical Physics
By A. Crisanti, G. Paladin, and A. Vulpiani

105 **Self-Trapped Excitons** By K. S. Song and R. T. Williams

106 **Physics of High-Temperature Superconductors** Editors: S. Maekawa and M. Sato

107 **Electronic Properties of Polymers** Orientation and Dimensionality of Conjugated Systems
Editors: H. Kuzmany, M. Mehring, and S. Roth

108 **Site Symmetry in Crystals** Theory and Applications
By R. A. Evarestov and V. P. Smirnov

109 **Transport Phenomena in Mesoscopic Systems** Editors: H. Fukuyama and T. Ando

110 **Symmetry and Optical Phenomena in Superlattices and Other Heterostructures**
By E. L. Ivchenko and G. E. Pikus

111 **Low-Dimensional Electronic Systems** New Concepts
Editors: G. Bauer, F. Kuchar, and H. Heinrich

112 **Phonon Scattering in Condensed Matter VII** Editors: M. Meissner and R. O. Pohl

113 **Electronic Properties of High-T_C Superconductors**
Editors: H. Kuzmany, M. Mehring, and J. Fink

114 **Interatomic Potential and Structural Stability** Editors: K. Terakura and H. Akai

K. Terakura H. Akai (Eds.)

Interatomic Potential and Structural Stability

Proceedings of the
15th Taniguchi Symposium
Kashikojima, Japan, October 19–23, 1992

With 92 Figures

Springer-Verlag
Berlin Heidelberg New York
London Paris Tokyo
Hong Kong Barcelona
Budapest

Professor Dr. Kiyoyuki Terakura

Institute for Solid State Physics, The University of Tokyo
Roppongi, Minato-ku, Tokyo 106, Japan

Professor Dr. Hisazumi Akai

Department of Physics, Nara Medical University
Kashihara, Nara 634, Japan

Series Editors:

Professor Dr., Dres. h. c. Manuel Cardona
Professor Dr., Dr. h. c. Peter Fulde
Professor Dr., Dr. h. c. Klaus von Klitzing
Professor Dr., Dres. h. c. Hans-Joachim Queisser

Max-Planck-Institut für Festkörperforschung, Heisenbergstrasse 1
D-70569 Stuttgart 80, Germany

Managing Editor:

Dr. Helmut K.V. Lotsch

Springer-Verlag, Tiergartenstrasse 17, D-69121 Heidelberg, Germany

ISBN-13:978-3-642-84970-1 e-ISBN-13:978-3-642-84968-8
DOI: 10.1007/978-3-642-84968-8

Library of Congress Cataloging-in-Publication Data. Interatomic potential and structural stability: proceedings of the 15th Taniguchi Symposium, Kashikojima, Japan, October 18-23, 1992 / K. Terakura, H. Akai (eds.) p. cm. – (Springer series in solid-state sciences ; 114) Includes bibliographical references and index.
1. Electronic structure–Congresses. 2. Intermolecular forces–Congresses. I. Terakura, K. (Kiyoyuki), 1942- . II. Akai, H. (Hisazumi), 1947- . III. Series. QC176.8.E4I45 1993 530.4'11–dc20 93-17953

Softcover reprint of the hardcover 1st edition 1993

Typesetting: Camera ready copy from the authors/editors
54/3140 – 5 4 3 2 1 0 – Printed on acid-free paper

Participants in the Taniguchi International Symposium

October 19–23, 1992 (at Kashikojima, Japan)

Seated (left to right)
T. Fujiwara, M. Aoki, F. Ducastelle, M. Kaburagi,
J. M. Sanchez, R. Kubo, A. R. Williams, K. Terakura,
A. Oshiyama

Second row (left to right)
R. M. Martin, Ms. Martin, Ms. Ducastelle, Ms. Akai,
Ms. Nørskov, J. K. Nørskov, D. Knab,
Ms. Louie, S. G. Louie

Third row (left to right)
T. Asada, T. Oguchi, H. Akai, R. Car, R. O. Jones,
H. Ishida, M. Takahashi, Y. Morikawa, T. Mohri

Preface

This volume contains the proceedings of the 15th Taniguchi Symposium on "Interatomic Potential and Structural Stability" held at Kashikojima, Mie-prefecture, Japan, during October 19–23, 1992.

Mr. Toyosaburo Taniguchi, who is the former President of Toyobo Co., Ltd., and one of the leading figures in the textile industry in Japan, has been supporting many academic activities in natural and human sciences for many years. The Taniguchi International Symposium is a project conceived by him for the purpose of promoting the international friendship among young scientists as well as contributing to the progress of basic sciences. It consists of 17 divisions covering selected fields. Each division will hold a small-size international meeting on a selected topic every year. The total number of participants is limited to about twenty including several invited speakers from abroad.

"Theory of Condensed Matter" is one of the divisions of the Taniguchi International Symposium. The Steering Committee of the division formed by R. Kubo (Chairman), J. Kanamori (Secretary), T. Matsubara, K. Yosida, and E. Hanamura selected the topic "Interatomic Potential and Structural Stability" for the symposium in 1992.

Elucidation of the structural properties is the first step towards the understanding of the materials from the microscopic point of view. The aim of the symposium was to have an opportunity for the exchange of up-to-date information of the theoretical aspects in this context among experts. The talks can be classified crudely into the following three categories.

1. Several important proposals have been made recently to the theoretical methods for calculating the interatomic potentials and forces acting on atoms starting from electronic-structure calculations: the Car-Parrinello method, the effective-medium method, the bond-order potential and so on. Attempts are being made to develop even more ambitious methods such as the order-N method. The present status of these methods as well as some applications to actual systems were discussed.

2. Once the interatomic potentials or the forces acting on atoms are obtained, it is necessary to search for stable structures or to study the phase stability at finite temperatures. For this purpose, the microscopic quantum mechanical calculations have to be combined with the statistical physics treatment. In this context, some presentations were made on the alloy phase diagrams and the phase transition on the Si(001) surface.

3. Most of the ab-initio electronic-structure calculations in condensed-matter physics are based on the local-density approximation (LDA) in the density-functional theory. Although LDA has been playing a very important role, several evidences that indicate the insufficiency of LDA have been accumulated at the same time. "Going beyond LDA" is therefore an important subject. Some attempts in this direction were also discussed.

The symposium was held at a pleasant hotel located at the seashore. We hope that all the participants enjoyed good food and drink, and sightseeing as well as the fruitful academic activities. We would like to express our sincere gratitude to the Taniguchi Foundation for supporting this very nice symposium.

Tokyo
Nara
April 1993

K. Terakura
H. Akai

Contents

Introduction

K. Terakura

Institute for Solid State Physics, University of Tokyo,
Roppongi, Minato-ku, Tokyo 106, Japan

The remarkable progress in theoretical condensed-matter physics depends on the development and improvement of the methodology and also on the rapid progress in modern high-performance computers. As for the latter, although the performance of a single processor is still being improved, parallel computing will become increasingly important. **Williams** discussed the parallel computing from various aspects. It may be interesting to cite some of the subsection titles from his article: "Improved Pain/Gain Ratio", "Cost/Performance Versus Performance", "Budget-Limited Versus Time-Limited Computing", "Parallelism will be Medium-Grained", "Shared Versus Distributed Memory". A unique idea about the order-N method in the electronic-structure calculation was also presented and its relation to parallel computing was discussed.

One of the most noticeable events during the last decade in the ab-initio methods for electronic-structure calculations is the introduction of the Car-Parrinello (CP) method. It provides a new efficient algorithm for solving the Kohn-Sham equation as well as a new viewpoint concerning the simultaneous optimization of atomic arrangements and electronic structures. Nevertheless, in order to deal with real complex systems, we often have to handle very large number of atoms. For such purposes, it is not possible and not economical to perform even the CP-type ab-initio calculations. It is inevitable to use some sort of classical interatomic potentials. However, for the many-atom potentials, the functional form itself is hard to determine empirically. Two promising approaches were discussed in the symposium.

One is the effective-medium theory (EMT). In the standard EMT so far, the basic idea is that the atom in a given environment can be recplaced with an atom in a uniform electron gas whose electron density is adjusted properly. With some correction terms, EMT has been applied successfully to several problems in the bulk and on the surface. Recent attempts for a more elaborate ab-initio EMT was discussed by **Nørskov**. The application of the method to various situations of Al turned out to be very successful. The surface energies, vacancy formation energy and phonon frequencies can be accurately reproduced.

The other is the bond-order potential originally proposed by Pettifor and extended by his coworkers. The formulation is based on the tight-binding Hückel theory, and the bond-order potential takes account of the angular dependence properly. **Aoki** discussed basic aspects of the bond-order potential, such as the exact many-atom expansion theorem and presented some applications.

Springer Series in Solid-State Sciences, Volume 114
Interatomic Potential and Structural Stability Eds.: K. Terakura and H. Akai

In most cases, the CP method utilized the merit of the plane-wave expansion in the estimation of the Hellmann-Feynman force. Weakness of the ionic pseudopotential is a prerequisite condition for the efficiency of the plane-wave expansion. The first low elements and transition-metal elements do not satisfy this condition. Some important progress has been made in order to facilitate the calculation for systems containing these elements. These are the ultra-soft pseudopotential method, the full-potential linear-augmented plane-wave (FLAPW) method, and the KKR method.

Oguchi gave a brief summary of the methodological aspects of the FLAPW type approach proposed by Solar and Williams. The merit of this particular approach is that the electronic charge density and, therefore, the potential are rigorously continuous on the muffin-tin sphere surface. The validity of the frozen-augmentation approximation for calculating the forces was also discussed. The usefulness of the method was demonstrated by several calculations including phonon frequencies of ferromagnetic iron.

On the other hand, **Akai** argued that the KKR method with the muffin-tin potential model can be used to calculate forces acting on atoms fairly accurately and efficiently with a suitable expression for the forces. He applied the method to the interstitial impurities of Li and N in Mg.

Oshiyama studied the electronic and atomic structures for defects and also impurity diffusion in Si. He adopted the conjugate-gradient method for solving the electronic structure as well as for optimizing the atomic structure. The unpaired electron associated with a negatively charged vacancy state V^- produces superhyperfine fields at the neighboring sites. The calculated hyperfine tensor agrees well with the experimental ENDOR data. As for the impurity diffusion, the vacancy-mediated process and the interstitial-mediated one have nearly the same activation energy and it is difficult to judge which is the major process. He pointed out that the two processes respond differently to the pressure and that this fact may be used to identify the major diffusion process.

The elements belonging to the III B to VI B families in the periodic table are in the transient region from non-metallic to metallic substances and form various allotropes. Small clusters formed by these elements have also characteristic features. **Jones** studied the structural trends in clusters composed of these elements systematically by using the CP method. Amorphous and liquid states of Se and P were also studied. Based on these studies, he emphasized that the bonding trends in amouphous and liquid phases are evident even in clusters with less than 10 atoms.

The order–disorder phase transition on the Si(001) surface was discussed by **Terakura**. The problem was treated by the combination of the CP method and the Monte Carlo (MG) simulation. The system is approximated by a model Ising spin system and the interaction parameters between neighbors are derived from the total energies for some arrangements of the asymmetric dimers estimated by the CP method. The MC simulation is then performed. It was demonstrated that the small amount of dimer defects on the surface produces a dramatic effect on the nature of the phase transition.

Fujiwara discussed the Al adsorption on Si surfaces: Si(111)–($\sqrt{3} \times \sqrt{3}$)R30°–Al and higher coverage regime of Al adsorption on the Si(001) surface. It was pointed out that the interlayer distance between Al and Si for the two Al(110) layers on the Si(001) surface is very large. This fact was attributed to the reason of epitaxial growth of Al(110) layers on the Si(001) surface without transition layers. Fujiwara also discussed a method for dealing with the metallic systems within the CP scheme. They introduced the fractional occupancy and treated the occupation number as a dynamical variable.

Morikawa presented detailed studies of the alkali-metal adsorption on the Si(001) surface by the CP method. The calculated coverage dependence of the adsorption energy gives a strong support to the double-layer model rather than the Levine model as the saturation coverage for Na nd K. The coverage of the former (latter) model is 1.0 (0.5) monolayers (ML). In contrast to Na and K, Li will be adsorbed beyond 1.0 ML at room temperature. The optimized structure for 2 ML of Li seems to suggest the Li–slicide formation.

The structural and vibrational properties of alkali-metal overlayers on metal surfaces were discussed by **Ishida**. He first emphasized the importance of the general notion that the formation of the adatom–adatom bond tends to weaken the adatom–substrate bond. The coverage dependence of the work function can be interpreted in terms of such a covalency effect rather than the conventional charge-transfer effect. He extended his theory to a study of the coverage dependence of the dynamical charge associated with the adatom vibration.

The discovery of the scanning tunneling microscopy (STM) opened a new era in surface science. It is particulary powerful for obtaining the local atomic and electronic structures. Despite its remarkable experimental progress, the reason of the atomic-scale resolution by using a tip with a curvature of several hundreds of Ångstroms has not been fully understood. **Tsukada** first demonstrated, taking the graphite surface as an example, that the atomic structure of the tip influences the STM image significantly. He then presented an analysis for the Si(111)–$\sqrt{3} \times \sqrt{3}$–Ag system. This is a typical example which shows the necessity of detailed theoretical analysis because the STM image does not correspond directly to the atomic arrangement on the surface. Some other systems are also discussed.

A general review on the order and phase stability of alloys was given by **Ducastelle**. Emphasis was placed on the tight-binding model which is particularly useful for the analysis of the transition-metal alloys. Compact discussions on the relation between moments of the density of states and the effective interatomic interactions are followed by generalized perturbation theory for deriving the interatomic interactions starting from the coherent-potential approximation for random alloys. Some examples in the structural trends in the actual transition-metal compounds are briefly discussed.

A first-principles theory for the alloy phase diagram was discussed by **Sanchez**. The ab-initio electronic total-energy calculation is combined with the statistical mechanics theory for estimating configurational entropy. The vibrational modes are also taken into account in the total free energy within the Debye-Grüneisen approximation. A brief review was given on a formal framework for the treatment of short-range order effects in the configurational energy (the Connolly and Williams method) and in the configurational entropy (the cluster variation method). The method was applied to three binary alloys in the Ru–Zr–Nb system with success.

With a given set of interatomic interaction parameters, it is still not easy to find out a ground-state atomic arrangement. **Kaburagi** discussed such kind of problems taking one-dimensional long-period structures as an example. In particular, he discussed the close-packed based family represented by a stacking sequence like ABAC.... The actual systems related to this family are pseudo-binary Laves phase $Mg(Ni_{1-x}Cu_x)_2$, interstitial O atoms in ZrO_x ($x \approx 1/3$), rare-earth metals, etc.

Kinetic transitions from one atomic configuration to another is an important subject in materials science. For instance, time dependent stability of a metastable state is a crucial factor for utilizing the metastable phase to realize a possible advanced material through non-equilibrium processes. **Mohri** studied the time evolution of short-range and long-range order parameters of an fcc-based spin system by using the path probability method (PPM) originally proposed by Kikuchi. Discussions are given on the order relaxation in the aging process, nucleation-growth ordering and spinodal ordering.

"Going beyond LDA" is an important subject for the first-principles electronic-structure calculations. While LDA has been playing an important role in a variety of materials, several evidences have also been accumulated which clearly indicate the insufficiency, and sometimes even the breakdown of LDA.

One straightforward extension of LDA is to take account of the electron density gradient corrections. Several test calculations were already made for some proposals related to the gradient corrections. Although there are some conflicting opinions about the virtue of the gradient corrections, **Asada** presented the detailed calculations for the 3d transition metals and concluded that the generalized gradient approximation (GGA) by Wang and Perdew and Perdew gives more sensible results than LDA definitely for the 3d transition metals. For example, the pressure-induced phase transition of Fe from the bcc ferromagnetic state to the hcp nonmagnetic state can be reproduced quite accurately by the GGA, while not at all by the LDA. The descrption of the structure and magnetism of Mn is also significantly improved by the GGA.

However, the GGA does not always give better results than the LDA and is not capable of describing strongly correlated systems. As a more elaborate treatment for the electron correlation, quantum Monte Carlo is regarded as a promising approach. **Louie** adopted a variational Monte-Carlo (VMC) technique in several problems, such as cohesive properties of bulk semiconductors, the magnetic-field-induced Wigner crystal in two dimensions and the magnetic properties of bcc hydrogen. These results clearly suggest the technique to have high reliability.

Martin compared the VMC with the diffusion Monte-Carlo (DMC) method. Although VMC is simpler than DMC computationally and conceptually, the result by VMC depends strongly on the trial wave function. On the other hand, DMC has the advantage of finding exact (or nearly exact) ground-state energies and correlations beyond those included in the trial wave function. A new approach to *core/valence partitioning* that leads to *quasiparticle pseudopotentials* for the many-body calculation for valence electrons was also discussed.

3d transition-metal compounds are typical systems where the electron correlation may play a crucial role and the elucidation of their electronic properties has been one of the central themes of condensed-matter physics. Here, NiS is a very interesting material which shows the metal–insulator transition by changing temperature and pressure. An obvious breakdown of the one-electron theory was reported by Fujimori et al. some years ago by comparing the band-structure calculation and the photoemission spectrum particularly in the metallic phase. **Takahashi** studied the problem first by using a simple linear-chain model combined with the exact diagonalization method and also a realistic three-dimensional model with the local-Ansatz approach proposed originally by Fulde.

Part I

New Developments in Methodology

Parallel Computing and Order-N Methods

A.R. Williams

IBM Watson Research Center, Yorktown Heights, NY 10598, USA

Abstract. Parallel computing will become increasingly important in the coming decade, its growth will be determined by the benefit obtained for the required recoding, which varies widely. The cost effectiveness of CMOS-RISC microprocessors will accelerate the growth of parallel computing by making more Unix processors available for parallel utilization. Algorithms well suited to the exploitation of the medium granularity of the parallel computers available in the 90s, can dramatically increase the utility of electronic-structure calculations for both science and materials engineering. This paper attempts to characterize the forces that will determine the type of computers available to us in this decade and presents an example of the type of algorithm which is well suited to such computers.

1. Parallel Computing in the 1990s

The 1990s will be quite different from the 1980s in the context of parallel computing. The principal developments underlying this difference are the performance and cost/performance of CMOS RISC microprocessors and the reduction of interprocessor communication latency to the microsecond range.

1.1 Improved Pain/Gain Ratio

An achievement with pervasive implications for computing, both parallel and serial, is the performance parity achieved by CMOS RISC microprocessors, relative to bipolar mainframe computers. Among the implications of this achievement is the fact that the pain/gain ratio for conversion to parallel computation has improved dramatically. That is, whereas in the 1980s parallel computers were composed of PC-class processors with a few megabytes of memory, the building blocks of new parallel computers are workstation-class processors, each with the power of a mainframe and each with the memory and storage of a mainframe. This, in turn, implies that there is *immediate* benefit to parallelization; the parallelism need not be "massive" to be beneficial. Additionally, the spreading of an application across multiple processors is simplified by both the power of RISC processors and the memory capacity that typically accompanies them. The 10-100 megaflop performance of these microprocessors means that scalar portions of an application no longer run at PC speeds. Memories on the scale of 100 megabytes mean that it is no longer necessary to spread all arrays across the parallel computer as a preliminary to execution, not to mention benefit. So, whereas in the 1980s parallel computing attracted primarily adventurers with an intrinsic interest in parallel computing, in the 1990s scientists and engineers for whom conversion to parallel computing has no intrinsic appeal, will be attracted by the competitive virtues of parallel computing.

1.2 Cost/Performance versus Performance

A second factor which will accelerate the spread of parallel computation in the 90s is the cost/performance advantage of CMOS RISC microprocessors. Workstations provide megaflops for a cost of several hundred dollars; mainframes provide them for several thousand. This cost advantage is independent of the performance parity mentioned above. The cost/performance advantage of the new processors means that existing budgets will tend to acquire a large *number* of processors. The *number* of processors is especially significant, because parallel computing offers little benefit until the number of processors exceeds the number of sharing users. Thus, for example, there is seldom an advantage to allocating the multiple processors of an expensive mainframe, such as a Cray or IBM 3090, to a single task. As clusters and "farms" of compute-server RISC workstations become common, the likelihood of the number of processors exceeding the number of sharing users increases, especially at night and on weekends.

Another manifestation of the great effectiveness of the CMOS-RISC combination is that cost/performance is no longer monotonic in cost; it is minimized by CMOS-RISC. That is, workstations are more expensive than PCs, but they are *much* more powerful (10-100 megaflops versus approximately one megaflop /1/). For this reason, CMOS-RISC microprocessors are the preferred building block for parallel computers, and this fact, in turn, drives the granularity of parallel computing.

1.3. Budget-limited versus Time-limited Computing

While the performance and cost/performance of CMOS microprocessors will encourage the growth of parallel computation, another fact of life will deter that growth. This is the fact that parallel computing does not increase anyone's computing budget; it merely provides a way of spending one's budget faster. So, to a first approximation, parallel computing does nothing for budget-limited computing; it benefits only time-limited computing, and the latter is relatively rare. Here again it is the gains in cost/performance derived from CMOS-RISC technology that will drive the expansion of parallel computing. So, while there are numerous technical challenges, such as low-latency communication and convenient programming languages, required for effective parallel computing; it is the perhaps less romantic challenges of cost/performance that are critical. For example, the much discussed goal of a teraflop computer is less a technical challenge (We can build one today.) than a cost challenge (We cannot afford the one we can build today.). In this regard, parallel computing is similar to supersonic commercial air travel.

1.4 Parallelism will be Medium-grained

The parallel computers of the 1990s will be medium-grained. The granularity of parallel computation is determined by the ratio of interprocessor communication latency to intraprocessor computation speed. The focussing of VLSI technology on the interprocessor communication problem is taking latency from milliseconds to microseconds, a tremendous technical achievement, but it remains true that arithmetic operations require only nanoseconds. Computation will remain two or three orders of magnitude faster than communication for the foreseeable future. This fundamental granularity-setting speed ratio is not a matter of the speed of light, the length of wires or network topologies; it is the result of decades of progress in the design of microprocessors, their pipelines, multiple arithmetic units, memory hierarchies, wide memory busses and compilers that exploit all of these. So, while it is quite true that there is much more data than there are independent threads of control, the number of data elements will greatly exceed the number of processors, that is, medium granularity.

1.5 Shared versus Distributed Memory

The term shared memory is used to describe both hardware and programming interfaces. Memory hardware will almost certainly be physically distributed. Sharing common hardware memory over a fast, wide bus reduces memory access time, but it is difficult to extend to more than ten (or a few tens of) processors. With the possible exception of the new computer from Kendall Square, scalably parallel computers can be characterized as distributed-memory architectures.

With regard to programming interface, there is probably a consensus opinion that scientists, and programmers generally, would prefer a shared-memory programming interface, that is, the freedom to formulate their application in a single, global address space. It is also probably a safe prediction that, on a time scale of only a few years, widely accepted programming interfaces, such as array-based generalizations of FORTRAN and C will be generally available. What is less clear is whether or not, or to what extent, scientists and engineers with a substantial investment in *existing* serial code will bite the bigger bullet of recoding their applications in parallel-but-shared-memory languages or accept the burden of adding message-passing constructs to their (dusty) serial code, or whether the pain of both paths will cause them to avoid parallel computation altogether. There will be examples of all three; the interesting question is what the distribution over these three possibilities will be. Message passing is undeniably the assembly language of parallel computing.

The historical precedent argues against the widespread use of message passing, but this portion of history was written when the investment in existing code was much smaller.

2 Electronic-Structure Calculations

Electronic-structure calculations offer insight into the behavior of chemically bonded systems of enormous variety, from ionic insulators to semiconductors to metals, including systems as complicated as those used for magneto-optics/2/. Specialized approximations and computational procedures have been developed to exploit properties of various systems. Most of these procedures have retained, however, a nonlinear dependence of computational effort on system size. The advances of this context from pseudopotentials, to augmented plane and spherical waves to the techniques introduced by Car and Parrinello/3/ have retained this nonlinear, cubic or worse, dependence on system size.

2.1 Order-N Methods

Inspired by Wilson's focus on "order N" as an objective, a series of recent developments has been aimed at reducing the dependence on system size to linear or N*ln(N). Most of these developments depend on the fact that density functional theory, despite its now well established accuracy, does not require the resolution of individual electronic states, but requires only a partitioning of the solution space into occupied and unoccupied subspaces.

2.2 Approaches taken by others

Recent efforts can be grouped into two areas, methods that work directly with the density matrix/4,5/, and methods exploiting a local approach to electronic structure/6,7/, such as those based on the recursion method. We present here preliminary results of an effort to exploit the concept of the renormalization group. Fundamental to the intuition underlying our approach is the tight-binding picture of bonding pioneered by Friedel, Pettifor and others. The basic idea is that it is only the interaction of states across the Fermi level, that is, interactions between occupied and unoccupied states that determine the strength of the chemical bond. This picture provides straightforward interpretations of the weakness of the interactions among rare-gas atoms and of the parabolic dependence of the cohesive energy as we proceed through each transition-metal series.

2.3 Renormalization-Group Approach

Consider a one-dimensional chain of hydrogen atoms. The minimal description of the electronic states of this system consists of expanding the individual electronic states of the chain in linear combinations of hydrogen 1s orbitals. Even in this "minimal", but chemically incisive description, the computational effort scales as the *cube* of the length of the chain.

Consider now an alternative approach to the same one-dimensional chain. Divide the chain into adjacent groups of four atoms. Diagonalize the Hamiltonian in each four-dimensional space, each diagonalization yielding four eigenstates. The energy diagonalization orders the subspace according to energy. This ordering allows us to partition the subspace into occupied and unoccupied subspaces. In each of these we now diagonalize an operator, such as x squared, thereby ordering the functions of the occupied and unoccupied subspaces according to their relative localization. As with conventional applications of the renormalization group/8/, we discard the most localized of the unoccupied states. In addition, and unlike conventional use of the renormalization group, we assume that the most localized member of is both fully occupied and fully localized. The contribution of this state is added to the integrated quantities of interest. Eliminating from further consideration two of the four states in each subspace allows us to combine the two remaining states in adjacent chain segments to once again form a four-state complex. We diagonalize the N/8 new four-dimensional repesentations of the Hamiltonian and x squared, and continue.

Only very preliminary numerical experiments have been carried out to assess the effectiveness of this approach. These experiments have considered only the simple problem stated above, a one-dimensional chain of hydrogen atoms. The size of the basic cluster which is iteratively formed and diagonalized is a fundamental parameter of the model. That is, we can, for example, start with groups of eight sites eliminating two states and occupying two states so as to have eight states at each level of the iteration. In addition, some of the interactions neglected for a given cluster size can be included perturbatively.

The principal fundamental problem with the technique is that the space of functions formed by the iteration is only approximately orthogonal. The coefficients in LCAO expansion of the states are nonzero only mutually exclusive subspaces (clusters), but nondiagonal overlap matrix elements provide a weak intercluster coupling. In the "toy" problems considered thus far, this effect has been quite small, but further study is required.

The technique is not restricted to LCAO representations, and could be applied, for example, to a discretized form of the Schrodinger equation.

2.4 Three-Dimensional Problems

The decomposition of three-dimensional problems into a binary tree appears to be a convenient and effective way in which to cast general problems in a form amenable to the technique described above. That is, the 3-D arrangement of atoms is partitioned into two smaller subsystems by a surface. Each substem is then itself partitioned, and so on. The interactions diagonalized at each stage of the iteration are those coupling through the partitions.

2.5 The Renormalization-Group and Parallel Computation

The calculations implied by the scheme described above are amenable to medium-grained parallel execution, but imperfectly. For the portion of the tree near the leaves, the number of subclusters will exceed the number of processors, permitting different branches of the tree to be handled by different processors. Parallelization in the portion of the tree in which the number of branches is smaller than the number of processors, will be more complicated, but certainly not impossible. It seems likely that convenience, budgets and machine availability will conspire to undermine the motivation for parallelizing the narrow portion of the tree. The medium granularity anticipated by the arguments earlier in this paper imply that this scheme is well suited to the machines on which it is likely to be used.

Acknowledgements

The electronic-structure portion of this work is the result of a collaboration with Ellen Stechel and Peter Feibelman of Sandia Laboratory in Albuquerque NM. The computer-science portion of the paper reflects the author's analysis and opinions, and should not be interpreted to be those of the IBM Corporation.

References

1. J. Dongarra, Computer Architecture News, Vol. 20, 22 (1992).
2. P.M. Oppeneer, T. Mauer, J. Sticht and J. Kubler, Phys. Rev. B, **45**, 10924 (1992).
3. R. Car and M. Parrinello, Phys. Rev. Lett., **55**, 2471 (1985).
4. D. Vanderbilt *private communication*
5. G. Galli and M. Parrinello, **in press**
6. W. Yang, Phys. Rev. Lett., **66**, 1438 (1992).
7. S. Baroni and P. Giannozzi, Europhys. Lett., **17**, 547 (1992).
8. S. R. White and R. M. Noack, Phys. Rev. Lett., **68**, 3487 (1992).

Ab Initio Effective-Medium Theory for Al

K. Stokbro, N. Chetty, K.W. Jacobsen, and J.K. Nørskov

Laboratory of Applied Physics, Technical University of Denmark, DK-2800 Lyngby, Denmark

A new method based on the effective-medium theory is presented for calculating the total energy for a system of interacting atoms. It is *ab initio* in the sense that all ingoing quantities are calculated directly without fitting to experiment or other calculations. It is approximate in the sense that simplifications in the total energy expression have been introduced and various numerical interpolation schemes have been used to get a method which is computationally efficient. The approximations all rely strongly on the variational properties of the total-energy functional. The method is tested for various bulk and surface properties of Al, and is found to be superior to the usual effective-medium theory and related methods.

1 Introduction

We present a method for total-energy calculations of solids for which the computational effort scales linearly with the number of atoms in the system. It is an *ab initio* method, there is no fitting to experiment or other calculations. The method is approximate and the level of accuracy can be chosen from very accurate and time consuming to more approximate and computationally efficient. It is only for the more approximate versions that the calculational effort scales linearly with the number of atoms in the system. At the most approximate level the method is identical to the usual effective-medium theory [1] and thereby mathematically similar to a number of frequently used and very efficient semi-empirical methods like the embedded-atom method [2], the corrected effective-medium theory [3], the Finnis-Sinclair [4] method, or the glue model [5].

The method has been derived in Ref. [6] and we will only briefly review the basic features here. The main purpose of the present paper is to present an implementation of the method for Al. A range of bulk and surface properties will be calculated using the method and compared to more time consuming methods and to experiments. We will show that even at a level of approximation where the method is extremely efficient it gives a very accurate description superior to that provided by the usual effective-medium theory and equivalent methods.

2 The *ab initio* effective-medium theory

The starting point is the density-functional theory of Hohenberg and Kohn [7], and we shall be working within the local-density approximation for exchange and correlation, although the theory may be generalized to more elaborate schemes for treating these effects. The total energy of a system of atoms is a unique functional $E[n]$ of the electron density $n(r)$ of the system, and $E[n]$ is stationary and variational around the ground-state electron density. Usually the ground-state density is determined by a self-consistent solution of the Kohn-Sham equations.

An equally good energy expression is the Harris functional [8, 9] which is stationary but not necessarily variational at the ground state energy [11, 12]. The Harris functional only depends on the input charge density, whereas the Kohn-Sham functional depends on an admixture of the input and output densities. It has been shown that overlapping free-atom densities constitutes a good input to the Harris functional for molecular systems and bulk systems where, in the latter, cohesive energies, lattice constants, phonon frequencies and bulk moduli can be accurately determined [13]. A well known problem using this ansatz for the input density, however, is the complete failure to determine surface energies [14].

In search for a better input density we first limit ourselves to the case where the total electron density can be written as a superposition of atom-like densities Δn_{atom}:

Springer Series in Solid-State Sciences, Volume 114
Interatomic Potential and Structural Stability Eds.: K. Terakura and H. Akai

Table 1: Calculated bulk and surface properties of Al in the FCC lattice using the selfconsistent (SC) and Harris (H) methods, the new parametrized Effective-Medium potential (EMT), and the simplest EMT from Ref. [1]. a)Nearest-neighbor distance (d_1^0), cohesive energy (E_c), and bulk modulus (B). Nottice in the new EMT the bulk FCC crystal has been taken as the reference system so that these quantities are exactly calculated, whereas in the simplest EMT of Ref. [1] an atom in a homogeneous electron gas has been used as the reference system so that these quantities are only approximate. b) Energy of formation of Al for the three principal unrelaxed surfaces and the ideal vacancy. The energies are per surface atom or per atom around the vacancy. The vacancy formation energy is then twelve (the number of nearest neighbors to the vacancy) times the vacancy number in the Table. c) The frequencies of the transverse and longitudinal X phonon, taken as an average of two frozen phonon calculations with displacements 0.02 and 0.03 in units of the lattice constant.

Quantity	SC	H	*newEMT*	Ref. [1]
a) BULK				
d_1^0(Bohr)	5.284	5.292	5.292	5.44
E_c (eV)	4.22	4.22	4.22	3.3
B (Mbar)	0.89	0.89	0.89	0.8
b) SURFACE and VACANCY				
(111) (eV)	0.43	0.46	0.43	0.31
(100) (eV)	0.52	0.55	0.62	0.42
(110) (eV)	0.80	0.86	0.86	0.65
vacancy	0.05	0.06	0.09	0.10
c) PHONONS				
X_L (THz)	9.4	9.1	9.0	7.5
X_T (THz)	5.5	5.2	5.2	5.3

$$n_{tot}(r) = \sum_{R_\mu} \Delta n_{atom}(r - R_\mu)\,, \tag{1}$$

where R_μ denote the position of the atomic nuclei. This is essential for an efficient method in particular in dynamical simulations where the atoms are allowed to move around. In reciprocal space this can be written

$$n_{tot}(G) = S(G)\Delta n_{atom}(G)\,, \tag{2}$$

where **G** is a reciprocal lattice vector, and $S(G)$ is the structure factor [15].

From a set of self-consistent calculations of $n_{tot}(G)$ for different structures and lattice constants it turns out that $\Delta n_{atom}(q)$ falls on an 'universal' curve which we term the optimized electron density[16]. Interpolating between the reciprocal space values where it is known and Fourier transforming back into real space gives an 'atomic' electron density, the 'optimized' atomic-electron density. Using this in conjunction with the ansatz Eq. (1) as input into the Harris functional gives excellent results for all cases studied as can be seen in Table 1. This is the first level of approximation.

In order to proceed, we introduce the concept of an effective medium or reference system. The idea is that for each atom i in the system, we look for a reference system where the atom is in 'similar' surroundings, but where the symmetry is higher so that the total-energy calculation is simpler. We then write the total energy of the system of interest

$$E = E + \sum_i e^{ref}(s_i) - \sum_i e^{ref}(s_i) \tag{3}$$

$$= \sum_i e^{ref}(s_i) + \Delta T + \Delta G. \tag{4}$$

Here $e^{ref}(s_i)$ is the energy of atom i in the reference system characterized by some parameter s_i, and the kinetic, and the electrostatic and exchange-correlation energy differences are given by

$$\Delta T = T - \sum_i t_i^{ref} = \sum_i [t_i - t_i^{ref}] \tag{5}$$

$$\Delta G = G - \sum_i g_i^{ref} = \sum_i [g_i - g_i^{ref}]. \tag{6}$$

The small e, t and g denote energies per atom (projected onto the Wigner-Seitz cell) whereas the capital E, T and G denote energies for the whole system.

In our work, we choose the effective medium or reference system to be an FCC crystal of an appropriate volume; An alternative could be the atom embedded in a homogeneous electron gas.

Now consider an arbitrary configuration of Al atoms. In order to find the reference system for any of these atoms, say atom i, we must specify the lattice constant of the reference FCC crystal. In principle any lattice constant will do, but obviously, some choices are more sensible than others. The most sensible ones are those where $|\Delta T|$ is small. The choice we make here is an FCC crystal which has the *same* neutral-sphere radius s_i as the atom i in the system under study:

$$\int_{s_i} dr\, n(r) = Z, \tag{7}$$

where $Z = 3.0$ corresponds to the total valence charge of atomic Al. This choice is the one that is made in the usual effective-medium theory, and ensures that the atom is situated in the same average electron density. With this choice of determining the reference volume, we ensure that there is no long-range Coulomb interaction between the atomic-spheres.

Given the ansatz optimized atom density $\Delta n_{op}(r)$, the electrostatic and the exchange-correlation energies of the system are exactly determined by first overlapping $\Delta n_{op}(r)$ in the configuration under investigation. The ΔG term can thus be determined exactly.

The Δt term is calculated within the atomic-sphere approximation ASA for overlapping *neutral* spheres. In the ASA we can write

$$t_i = \sum_\alpha \epsilon_\alpha[v_{eff}]|_i - \int_{s_i} dr\, \bar{v}_{eff}(r)n(r) \tag{8}$$

$$\approx \sum_\alpha \epsilon_\alpha[v_{eff}^{ref}]|_i - \int_{s_i} dr\, \bar{v}_{eff}^{ref}(r)n(r), \tag{9}$$

where the sum over the eigenvalues is projected on the sphere at site i. In the second equation we have exploited the variational property of the kinetic energy with respect to changes in the effective potential. Notice that in Eq (9) only the last term solely depends on the potential in cell i, whereas the eigenvalue sum depends on the potential in the surroundings, too.

The difference in the kinetic energies between the atom in question and the atom in its associated reference system is now given in the ASA by

$$\Delta t = \sum_\alpha \epsilon_\alpha[\bar{v}_{eff}^{ref}]|_i - \sum_\alpha \epsilon_\alpha[\bar{v}_{eff}^{ref}] \tag{10}$$

$$- \int_{s_i} dr\, \bar{v}_{eff}^{ref}(r)[n(r) - n^{ref}(r)] \tag{11}$$

$$\equiv \Delta\epsilon + \Delta v. \tag{12}$$

The first term $\Delta\epsilon$, is the socalled one-electron correction term. The second term Δv can again be calculated exactly given the electron density ansatz. If we also calculated the Δg term within the ASA we would get $\Delta g = -\Delta v$, since $\delta g/\delta n = v_{eff}$, to first order in $n - n^{ref}$. The sum of the two terms we therefore call the atomic-sphere correction:

$$\Delta E_{AS} = \sum_i [\Delta g_i + \Delta v_i]. \tag{13}$$

Up to now we have made two sets of approximations. First we have used the optimized-density ansatz. This has been tested and works extremely well. The reason is the variational property of the total energy functional. The second approximation is the ASA for the kinetic-energy functional. We have also tested this by calculating $\Delta\epsilon$ in the ASA using the tight binding LMTO scheme as described in Ref. [6]. The result is shown in the second column of Table 2. All potential parameters and structure constants have been calculated and the tight-binding hamiltonian is solved using the recursion method. The second moment of the density of states is the only quantity in the tight binding problem that has the form of a pair sum like the average electron density in the full problem. We therefore choose the reference system to be the FCC crystal with the same second moment, in the tight binding evaluation of $\Delta\epsilon$. In Table 2 we also show the result of a recursion calculation keeping only terms up to and including the fourth moment. Clearly a fourth moment expansion is sufficient for convergence of the one- electron energy *difference*, even though the individual contributions are far from converged at this level of approximation. It is this property of the the present method that makes it scale linearly with the number of atoms in the system.

Table 2: The one electron correction part of the total energies of the three principal surfaces and the vacancy considered in part b) of Table 1. The Table shows the correction obtained from the Harris construction together with result of a LMTO tight-binding scheme using different levels of approximations. The correction TB_1 is obtained from a sp model taken the recursion method to convergence. The correction TB_2 is obtained from a sp model taken the recursion method only to the second level, which is similar to only including moments upto the fourth moment. The correction TB_3 is obtained from the one band model constructed in section 3.

Quantity	E_{1BL}	TB_1	TB_2	TB_3
SURFACE and VACANCY				
(111) (eV)	-0.13	-0.13	-0.15	-0.15
(100) (eV)	-0.20	-0.22	-0.15	-0.10
(110) (eV)	-0.23	-0.29	-0.23	-0.24
vacancy	-0.11	-0.05	-0.08	-0.09

3 The Parametrized Effective-Medium potential

In this section we use the *ab initio* calculations to extract a parametrized version of the effective-medium potential, in order to obtain a simple and efficient scheme. Even at this level of approximation we get good quantitative results as shown in the third column of Table 1.

We start by extracting the following relations

$$s = d/\beta + \alpha \tag{14}$$
$$\bar{n}(s) = \bar{n}_0 \, e^{-\eta s} \tag{15}$$
$$\Delta\bar{n}(s,r) = \Delta\bar{n}_0 \, e^{\eta_1 s - \eta_2 (r/\beta + \alpha)}, \tag{16}$$

from the calculated optimized density. These are: A linear dependence between the neutral-sphere radius s and the nearest-neighbor distance d in the FCC structure; An exponential form between the average density $\bar{n}$ and the neutral-sphere radius; And the average-density contribution $\Delta\bar{n}$ from a neighbor to be given as a product of exponentials of the neutral-sphere radius and the distance to the neighbor. From the relation $\bar{n}(s_i) = \sum_j \Delta\bar{n}(s_i, r_{ij})$ we get a constrain on the parameters $\eta_2 = \eta + \eta_1$ and $\Delta\bar{n}_0 = \bar{n}_0/12$, if only nearest-neighbor contributions are included. Left are the parameters α, β, η and η_1, which we find by making a least squares fit in the parameter range of interest. The values are $0.1536, 1.9436, 2.449$ and 0.836, respectively, where the unit length is one bohr radius.

We are now able to calculate the neutral-sphere radius for an arbitrary system by making a simple sum over the neighbor distances

$$s_i \simeq \frac{1}{\eta_2} \ln\left[\frac{1}{12\gamma(d_0)} \sum_j e^{-\eta_2 (r_{ij}/\beta' + \alpha')}\right]. \tag{17}$$

In order to include more than nearest neighbors we have introduced the function

$$\gamma(d) = \frac{1}{12} e^{-\eta_2 (d/\beta' + \alpha')} \sum_j e^{-\eta_2 (r_{ij}^{ref}/\beta' + \alpha')} \tag{18}$$

given by the sum over the neighbors in the FCC reference system within the cutoff radius. If we expand Eq. (17) to first order in d we get the relation $s = (1/\beta' - \delta)d + (\alpha' + \delta d_0)$, in terms of the small parameter $\delta = \gamma'(d_0)/(\eta_2 \gamma(d_0))$. Comparison with Eq. (14) gives a selfconsistent equation for β' and α' which can be solved once and for all. In column one and two of Table 3 we show the neutral-sphere radius calculated from Eq. (17) for various situations using a cutoff between the third and fourth nearest-neighbor shells, compared with the correct results.

We are now able to calculate the energy of the reference system by parametrizing the binding energy of the FCC reference system as a function of the neutral-sphere radius. In terms of the function $B(x) = (1 + x)\exp(-x)$ [10] this energy is given by the cohesive function

$$E_C \simeq \sum_i E_0 B(\lambda(s_i - s_0)), \tag{19}$$

Table 3: The neutral-sphere radius and the three energy parts of the effective-medium potential for the structures considered in Table 1. The neutral-sphere radius quoted for each surface is that of the outermost atom, and for the vacancy the neutral-sphere radius of an atom next to the vacancy. For the phonons we quote the energies and neutral-sphere radii of a displacement of 0.03 in units of the lattice constant. For each quantity we compare the result obtained from the optimized density construction, with the result obtained from the parametrization in section 3.

Quantity	s	Eq. (17)	E_C	Eq. (19)	E_{AS}	Eq. (20)	E_{1EL}	Eq. (25)
a) SURFACE and VACANCY								
(111) (eV)	2.960	2.960	0.04	0.04	0.55	0.55	-0.13	-0.15
(100) (eV)	2.996	2.993	0.08	0.07	0.67	0.65	-0.20	-0.10
(110) (eV)	3.036	3.034	0.13	0.12	0.96	0.97	-0.23	-0.24
vacancy	2.897	2.897	0.005	0.003	0.161	0.176	-0.110	-0.094
b) PHONONS								
X_L (eV)	2.854	2.846	0.003	0.003	0.182	0.167	-0.121	-0.107
X_T (eV)	2.865	2.861	0.001	0.000	0.077	0.077	-0.056	-0.056

where E_0 is the cohesive energy and λ is determined by the bulk modulus. The third and fourth column of Table 3 shows the quality of this approximation.

Next we parametrize the atomic-sphere correction ΔE_{AS} by writing it as a difference between a sum of pair-potential terms in the real and in the reference systems:

$$\Delta E_{AS} \simeq \sum_i [\sum_{j \neq i} V(r_{ij}) - \sum_{j \neq i} V(r_{ij}^{ref})] \tag{20}$$

Here $V(r)$ is the pair potential, and r_{ij}^{ref} are the interatomic distances in the reference FCC lattice where the the nearest-neighbor distance is $\beta(s_i - \alpha)$. This is exact for the electrostatic part of the energy and obviously holds very well even for the non-electrostatic parts as can be seen in the fifth and sixth column of Table 3. For this calculation we have taken an exponential form for the pair potential $V(r) = V_0 \exp(-\kappa r)$, and fixed the two parameters to give the correct values for the (111) surface and the transverse X phonon.

We now introduce a simpler tight-binding model for the one-electron correction, by describing the four Al bands (one s band and three p bands) as four degenerate bands. The motivation for this approximation is that the $S_{ss\sigma}$, $S_{sp\sigma}$, and $S_{pp\sigma}$ elements of the LMTO-structure constants have almost the same distance dependence [18]. The $S_{pp\pi}$ element have a slightly stronger distance dependence, but in the FCC structure its magnitude is a factor 6 smaller than the $S_{pp\sigma}$ element, so it only has a minor influence on the energies. This means that the one-band model will describe changes in coordination number and neighbor distances relative to the reference system excellently, and since these are the quantities which have the largest influence on the one-electron correction, we expect the model to work well in most situations. The places where the one-band model will fail are for structures like a stacking fault, in which the coordination numbers and neighbor distances are similar to those of the FCC system, and thus the main contribution to the one-electron correction stems from the difference in the angles between the bonds. The simplified hamiltonian we use can be written

$$h_{ij} = \bar{\Delta}_i^{\frac{1}{2}} \bar{S}_{ij} \bar{\Delta}_j^{\frac{1}{2}} \tag{21}$$

where $\bar{S}$ is the effective-structure matrix and $\bar{\Delta}$ the effective-potential parameter.

We will now try to make a parametrization of $\bar{S}$ and $\bar{\Delta}$ as a function of the neutral-sphere radii and the interatomic distances. If we are only changing the lattice constant of the system, then the structure matrix should be unchanged, so all the dependence of the lattice constant should be in the potential parameter. According to Ref. [17] Δ is approximately inversely proportional to the lattice constant cubed and therefore we take $\bar{\Delta} = \Delta_0/(d_i^{ref} d_j^{ref})^{\frac{3}{2}}$

We are now going to parametrize the structural dependence of the hamiltonian as a function of the interatomic distance relative to the neutral-sphere radii $\bar{r}_{ij} = r_{ij}/(s_i s_j)^{\frac{1}{2}}$, since this parameter does not depend upon the lattice constant. The structural dependence of the hamiltonian consists of two terms, the structure matrix and the dependence of Δ_0 upon structure. For the first term we take the $S_{ss\sigma}$ element of the structure matrix, which we calculate from the LMTO Dyson equation, using the most

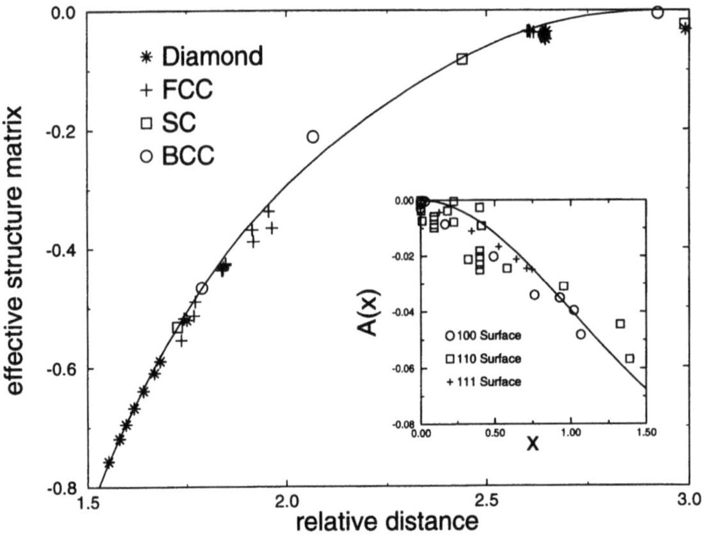

Figure 1: The structural part of the hamiltonian calculated using the LMTO dysons equation as a function of the relative distance between the atoms. The inlet shows the values which the asymmetri function $A(x)$ (defined in Eq. (24)) should take in order to get the correct values of the structure matrix at the three principal surfaces. The different data points were calculated from a cleaved FCC crystal with various distances between the two exposed surfaces.

screened sp representation[18]. The second term we estimate by calculating Δ_0 in a free electron model. In Figure 1 we have plotted $\bar{S} = \Delta_0/\Delta_0^{fcc}S_{ss\sigma}$ agains $\bar{r}$. As seen from the figure the points fall almost on a universal line which can be parametrized well by the functional form

$$\bar{S}(\bar{r}) \simeq S_0 \frac{e^{-\lambda \bar{r}}}{1+e^{a(\bar{r}-b)}}, \tag{22}$$

with the values $S_0 = -16.4, \lambda = 1.92, a = 6.0$, and $b = 2.6$ in units of bohr radius.

The structures used for the above interpolation formula are all very symmetric, in the sence that all atoms are equivalent. If one tries to calculate the structure matrix for a more unsymmetric case like a surface the correct value of the structure matrix is slightly higher than the one obtained from the interpolation formula. The reason is simply that the matrix elements are less screened at the surface because of the lack of symmetry. We describe the degree of symmetry by the parameter

$$x_i = \frac{|\sum_j w_{ij}\vec{r}_{ij}|}{(\sum_j w_{ij}^2 r_{ij}^2)^{\frac{1}{2}}} \tag{23}$$

where w_{ij} is some weight factor. We now parametrize the structure matrix as a function of this parameter

$$\bar{S}_{ij} = \bar{S}[\bar{r}_{ij}\,(1 + A(x_i x_j))], \tag{24}$$

where we have introduced the function A. In a symmetric structure we will have $x = 0$, so we constrain $A(0) = 0$ in order to retain the old fit in this case. We have found that if we choose the weight factor in Eq.(23) to be $w_{ij} = \exp(-1.27r_{ij})$ it is indeed possible to make a good parametrization of the asymmetry. In the inlet of Figure 1 we show the value $A(x)$ should take in order to obtain the correct value of the structure matrix from the interpolation formula at the three principal surfaces. As seen from the figure the points can be fitted well by the function, $A(x) = -0.1x(1 - e^{-x^2/2})$.

The strength of the one-band model Δ_0^{fcc} we fix to the correct one-electron correction for the transverse X phonon. This choice makes the second moment of the one band model ten percent larger than that of the sp model. This is within the uncertainty of the potential parameters arising from the choice of the energy of linearizing in the LMTO formalism.

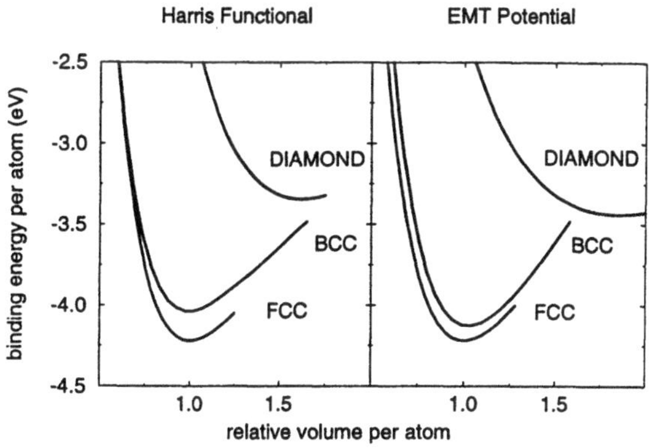

Figure 2: The total energy of the Aluminium FCC, BCC and Diamond phases, as a function of the volume per atom. The left graph shows the result obtained from the harris functional, while the right graph is the result of the parametrized effective-medium potential.

As in the sp model we solve the tight-binding hamiltonian using the recursion method terminating the chain at the second level, which corresponds to only including terms up to the fourth moment. For the termination we use the method of Nex [19]. The final form of the one-electron correction then becomes

$$\Delta E_{1el} = \sum_i \sqrt{\mu_2(i)}(\mathcal{E}_{nex}(a_i, b_i) - \mathcal{E}_{nex}(a^{fcc}, b^{fcc})). \quad (25)$$

The parameters are $a = \gamma_3$, $b = \sqrt{\gamma_4 - \gamma_3^2 - 1}$, and $\gamma_n = \mu_n/\mu_2^{n/2}$, where μ_n is the n'th moment of the density of states. The function $\mathcal{E}_{nex}$ is a second order polynomial which fits the energies obtained from the method of Nex in the parameter range of relevance. The last two columns of Table 3 shows the accuracy of this parametrization of the one-electron correction.

In Table 1 we have summerized the total energies obtained from using this parametrized version of the effective-medium potential. In all cases we get very good agreement with the self-consistent results and the results using the Harris functional.

Notice that if we also include the one-electron correction in the pair potential difference Eq. (20) then we get exactly the simplest form of the usual effective-medium theory. Table 1 also includes the results for Al at this level of approximation from Ref. [1]. The result shows the usual problem for this kind of approximation[2] that the absolute values of the surface energies are too small (whereas relative surface energies are much better); The anomalous lack of a $\sqrt{2}$ ratio for aluminium between the longitudinal and transversal zone-boundary phonons is not reproduced. The new *ab initio* effective-medium theory has solved these problems.

We have also used the potential to calculate the phase diagram of Aluminium in the FCC, BCC and Diamond structures. For comparison we have also calculated these energies using the optimized densities as input to the Harris functional. The result is shown in Figure 2. The FCC phase is reproduced almost exactly, which is simply due to the use of FCC as the reference system, such that the energy is determined, solely by the cohesive function. Also for the BCC and Diamond phases the cohesive function gives the leading term such that the equilibrium neutral-sphere radius in these phases are very close to s_0.

4 Conclusions

In conclusion, we have introduced a new *ab initio* effective-medium theory for total energy calculations in solids. It is computationally very efficient, and the computational effort scales linearly with the number of atoms in the system. We have tested the accuracy of the method by calculating various bulk and surface properties of Al, and shown that the self-consistent and Harris functional results are well reproduced. In order to obtain the correct structure matrix at the surface we introduced an asymmetry factor in the parametrization. Work is in progress to test whether this form is general enough to include other structures as well.

5 Acknowledgments

We are grateful to K. Kunc, O.H. Nielsen, R.J. Needs and R.M. Martin whose solid state programs we have used, and to E.L. Shirley who developed the pseudopotential routines. This work was in part supported by the Danish Research Councils through the Centre for Surface Reactivity.

References

[1] K.W. Jacobsen, J.K. Nørskov and M.J. Puska, Phys. Rev. B**35** (14), 7423 (1987).

[2] M.S. Daw and M.I. Baskes, Phys. Rev. Lett. **50**, 1285 (1983).

[3] S. B. Sinnot, M. S. Stave, T. J. Raeker, and A. E. DePristo, Phys. Rev. B **44**, 8927 (1991).

[4] M.W. Finnis and J.E. Sinclair, Phil. Mag. A **50** 45 (1984).

[5] F. Ercolessi, E. Tosatti, and M. Parinello, Phys. Rev. Lett. **57** , 719 (1986).

[6] N. Chetty, K. Stokbro,, K. W. Jacobsen, and J. K. Nørskov, Phys. Rev. B **46**, 3798 (1992).

[7] P. Hohenberg and W. Kohn, Phys. Rev. **136**, 3864 (1964), W. Kohn and L.J. Sham, Phys. Rev. **140** (4A), A1133 (1965).

[8] J. Harris, Phys. Rev. B**31** (4), 1770 (1985).

[9] W.M.C. Foulkes and R. Haydock, Phys. Rev. B**39** (17), 12520 (1989).

[10] J.H. Rose, J. Ferrante, and J.R. Smith, Phys. Rev. Lett. **47** , 675 (1981).

[11] E. Zaremba, J. Phys. Condens. Matter **2**, 2479 (1990).

[12] I.J. Robertson and B. Farid, Phys. Rev. Lett. **66** 3265 (1991).

[13] H.M. Polatoglou and M. Methfessel, Phys. Rev B**37** (17), 10403 (1988).

[14] A.J. Read and R.J. Needs, J. Phys. Condens. Matter **1**, 7565 (1989).

[15] J. Ihm, A. Zunger and M.L. Cohen, J. Phys. C**12**, 3792 (1979).

[16] N. Chetty, K.W. Jacobsen and J.K. Nørskov, Lett. J. Phys., Condens. Matter **3**, 5437 (1991).

[17] O.K. Andersen, O. Jepsen, and D. Glötzel, in *Highlights of Condensed-Matter Theory*, edited by F. Bassani, F. Fumi, and M. P. Tosi (North-Holland, New York, 1985).

[18] O.K. Andersen, O. Jepsen, and M. Sob, in *Electronic Bandstructure and its applications*, edited by M. Yussouff (Springer Lecture Notes, 1987).

[19] C.M.M. Nex, J. Phys. A. **11**, 653 (1978).

Angular-Dependent Many-Atom Bond-Order Potentials

M. Aoki[1,2], *P. Gumbsch*[3,4], *and D.G. Pettifor*[1]

[1]Department of Materials, University of Oxford,
Parks Road, Oxfor OX1 3PH, UK
[2]Department of Physics, Gifu University, Yanagido, Gifu 501-11, Japan
[3]Department of Mathematics, Imperial College, London SW7 2BZ, UK
[4]Max-Planck-Institut für Metallforschung,
Seestr. 92, W-7000 Stuttgart, Germany

Abstract. The state-of-the-art many-atom potentials for the bond order, which are derived within tight binding Hückel theory, are shown to be promising tools in modeling of the structural properties of materials. Explicit angular dependence of these potentials is naturally derived on the theoretical basis of the exact many-atom expansion theorem for the bond order. It is shown that a good convergence is achieved by satisfying a sum rule for the intersite Green's function.

1. Introduction

Computing-based theoretical studies in materials science have recently become more intensive and extensive. Atomistic simulation [1], which is one of the most promising approaches for modeling of materials, can be used to explain and also to predict structural stability and related properties of a materials. However, in this kind of study, it is essential to have the interatomic potentials which possesses both the fortes of simplicity and reliability. Although a treatment using simplified interatomic potential with an analytic functional form, of course, has to be less accurate than a fully first principles calculation, it will allow us to deal with a great number of atoms in unit cell. Moreover, it will provide us some room to interpret and model the properties of a material.

For covalent systems, the presence of angular forces is known to stabilize the diamond-type open structure of semiconductors and hcp→bcc→hcp→fcc structures through the non-magnetic transition metals series. Thus, simplest pairwise potentials and other angularly independent potentials [2, 3] are not appropriate for describing the bonding in these materials. Recently, many studies have been devoted to develop angularly dependent effective interatomic potentials [4, 5, 6, 7].

In this paper, we explain the formalism of angularly-dependent many-atom bond order potentials derived within tight binding Hückel theory. Section 2 introduces the formalism based on the theorem of exact many-atom expansion. In section 3, we establish sum rules for our bond order potentials and develop simplified approximate models. Then, we summarize in section 4.

Springer Series in Solid-State Sciences, Volume 114
Interatomic Potential and Structural Stability   Eds.: K. Terakura and H. Akai

2. Formalism

2.1 The bond order and bond energy within tight binding Hückel theory

The bond order can be defined clearly within tight binding Hückel theory as Coulson [8] proposed to predict bond lengths in molecules. For the dimer of two equivalent s-valent atoms i and j with the bond integral $H_{i,j}(< 0)$, we find bonding $|+\rangle$ and antibonding $|-\rangle$ states with energies $\pm H_{i,j}$, where $|\pm\rangle \equiv (|i\rangle \pm |j\rangle)/\sqrt{2}$ and $|i(j)\rangle$ are the atomic s orbitals. When the bonding and antibonding states are occupied by N^+ and N^- electrons per spin, respectively, the covalent bond energy of the dimer is counted to be

$$U_{\text{bond}}^{i,j} = 2H_{i,j}\Theta_{i,j} \tag{1}$$

with the bond order defined by

$$\Theta_{i,j} = N^+ - N^-. \tag{2}$$

The relationship Eq. (1) keeps its form even when the dimer is embedded in a surrounding atomic matrix and the influence of the local atomic environment enters the bond order. The total bond energy of a whole system is written

$$U_{\text{bond}} = \frac{1}{2}\sum_{i\neq j} U_{\text{bond}}^{i,j} \,. \tag{3}$$

The total binding energy is given by including an empirical pairwise repulsive contribution $\frac{1}{2}\sum_{i\neq j}\phi_{\text{rep}}^{i,j}(R_{ij})$ from the electrostatic and exchange-correlation energy of overlapping atomic charge densities [9].

Since $N^{\pm}$ are calculated by integrating the imaginary part of bonding / antibonding Green's functions $G_{00}^{\pm}(E) = \langle \pm|(E-H)^{-1}|\pm\rangle$, we get

$$\Theta_{i,j} = -\frac{2}{\pi}\text{Im}\int^{E_F} G_{i,j}(E+i0)dE, \tag{4}$$

where $G_{i,j}(E) = \langle i|(E-H)^{-1}|j\rangle$ are the intersite Green's functions, E_F is the Fermi energy and Im is the imaginary part.

Figure 1 schematically shows the bond order as a function of the number of electrons per bond per spin $N = N^+ + N^-$ for an isolated dimer and an

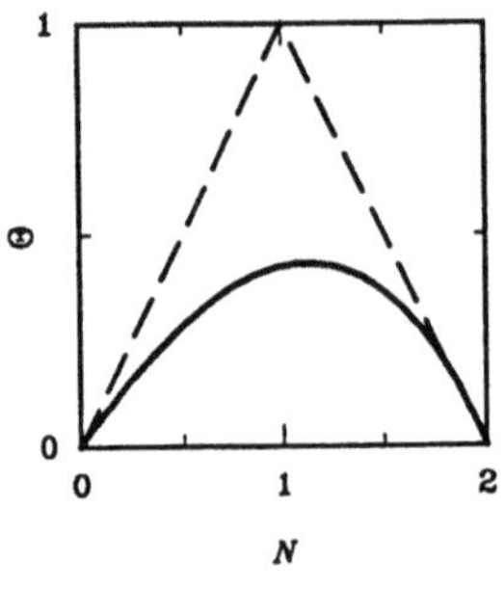

Figure 1. Schematic curves of bond order for isolated dimer (dashed) and embedded dimer (solid).

embedded dimer. The bond order of the embedded dimer displays some reduction, which depends on the local atomic coordination and configuration about a bond.

2.2 Theorem of exact many-atom expansion

To derive a basal theorem for our purpose, we prepare the Green's function

$$G_{00}^{\lambda}(E) = \langle u_0^{\lambda}|(E-H)^{-1}|u_0^{\lambda}\rangle, \tag{5}$$

with the state vector defined by

$$|u_0^{\lambda}\rangle = \frac{1}{\sqrt{2}}\left[|i\rangle + e^{i\theta}|j\rangle\right] \tag{6}$$

with $\theta = \cos^{-1}\lambda$, where λ $(-1 \leq \lambda \leq 1)$ is parameter to specify a state. The tight binding Hamiltonian H can be tri-diagonalized in diagonal elements $a_0^{\lambda}, a_1^{\lambda}, \cdots$ and off-diagonal elements $b_1^{\lambda}, b_2^{\lambda}, \cdots$ applying the Lanczos recursion algorithm [10]

$$b_1^{\lambda}|u_1^{\lambda}\rangle = H|u_0^{\lambda}\rangle - a_0^{\lambda}|u_0^{\lambda}\rangle, \tag{7}$$

$$b_2^{\lambda}|u_2^{\lambda}\rangle = H|u_1^{\lambda}\rangle - a_1^{\lambda}|u_1^{\lambda}\rangle - b_1^{\lambda}|u_0^{\lambda}\rangle, \tag{8}$$

$$b_3^{\lambda}|u_3^{\lambda}\rangle = H|u_2^{\lambda}\rangle - a_2^{\lambda}|u_2^{\lambda}\rangle - b_2^{\lambda}|u_1^{\lambda}\rangle \tag{9}$$

and so on, with the orthogonality condition $\langle u_m^{\lambda}|u_n^{\lambda}\rangle = \delta_{mn}$. It is well-known [11, 12] that the Green's function Eq. (5) can be represented in the continued fraction

$$G_{00}^{\lambda}(E) = \cfrac{1}{E - a_0^{\lambda} - \cfrac{(b_1^{\lambda})^2}{E - a_1^{\lambda} - \cfrac{(b_2^{\lambda})^2}{E - a_2^{\lambda} - \cdots}}}. \tag{10}$$

We find from Eqs. (5) and (6) that

$$G_{00}^{\lambda}(E) = G_{00}^{0}(E) + \lambda G_{i,j}(E) \tag{11}$$

is exactly linear with respect to λ and the same is Eq. (10). Therefore, from this linearity, we can develop many-atom expansion for the intersite Green's function, namely

$$G_{i,j}(E) = \left(\frac{dG_{00}^{\lambda}}{d\lambda}\right)_0 = \sum_{n=0}^{\infty}\left(\frac{\partial G_{00}^{\lambda}}{\partial a_n^{\lambda}}\right)_0 \delta a_n + \sum_{n=1}^{\infty}\left(\frac{\partial G_{00}^{\lambda}}{\partial b_n^{\lambda}}\right)_0 \delta b_n, \tag{12}$$

where $(\cdots)_0$ denotes that $\lambda \to 0$, and

$$\delta a_n \equiv \left(\frac{da_n^{\lambda}}{d\lambda}\right)_0 \quad \text{and} \quad \delta b_n \equiv \left(\frac{db_n^{\lambda}}{d\lambda}\right)_0. \tag{13}$$

The series representation in Eq. (12) is not an approximation to first order but it is *exact*.

It follows that the bond order, Eq. (4), takes the form of series expansion

$$\Theta_{i,j} = -2\left[\sum_{n=0}^{\infty} \chi_{0n,n0}(E_F)\delta a_n + \sum_{n=1}^{\infty} \chi_{0(n-1),n0}(E_F)2\delta b_n\right], \tag{14}$$

where the response functions $\chi_{0m,n0}(E_F)$ are defined by

$$\chi_{0m,n0}(E_F) = \frac{1}{\pi}\mathrm{Im}\int^{E_F} G^0_{0m}(E+i0)G^0_{n0}(E+i0)dE \tag{15}$$

with the off-diagonal Green's functions defined by $G^0_{0n}(E) = \langle u^0_0|(E-H)^{-1}|u^0_n\rangle$. Analytic expressions for these response functions has been obtained using recursion coefficients to a desired level [13].

It should be noted that bond order potentials within the present formalism, unlike perturbatively derived many-atom potentials [6, 7], do not need to have any reference system. This fact will be important when treating a system with a wide range of atomic configuration.

2.3 Moment and interference term

Explicit real space functional forms of the influence of the local atomic environment about a bond may be displayed by the moment (the average nth moment)

$$\mu_n = \frac{1}{2}[\langle i|H^n|i\rangle + \langle j|H^n|j\rangle] \tag{16}$$

and the interference term

$$\zeta_{n+1} = \langle i|H^n|j\rangle, \tag{17}$$

which can be expressed analytically using the bond angles θ_k and dihedral angles ϕ_k [14]. It is the well-known relationship between the moments and the recursion coefficients that

$$a^0_0 = \mu_1 \equiv 0, \tag{18}$$
$$(b^0_1)^2 = \mu_2, \tag{19}$$
$$a^0_1 = \mu_3/\mu_2, \tag{20}$$
$$(b^0_2)^2 = \mu_4/\mu_2 - \mu_3^2/\mu_2^2 - \mu_2, \tag{21}$$

where $\mu_1 = (E_i + E_j)/2 = 0$, with atomic orbital energy $E_{i(j)}$, is taken as the reference energy. Also, the differential coefficients δa_n and δb_n are calculated,

$$\delta a_0 = \zeta_2, \tag{22}$$
$$2\delta b_1 = \zeta_3/b^0_1, \tag{23}$$
$$\delta a_1 = \left[\zeta_4 - a^0_1\zeta_3 - 2(b^0_1)^2\zeta_2\right]/(b^0_1)^2, \tag{24}$$

$$
\begin{aligned}
2\delta b_2 &= \{\zeta_5 - 2a_1^0\zeta_4 - [2(b_1^0)^2 + (b_2^0)^2 - (a_1^0)^2]\zeta_3 \\
&\quad +2a_1^0(b_1^0)^2\zeta_2\}/\left[(b_1^0)^2 b_2^0\right], \qquad (25)
\end{aligned}
$$

It should be noted that δa_n and δb_n can also be given by linearizing $\Delta a_n = \frac{1}{2}(a_n^+ - a_n^-)$ and $\Delta b_n = \frac{1}{2}(b_n^+ - b_n^-)$ to first order in any ζ_k, using the recursion coefficients $\{a_n^\pm, b_n^\pm\}$ corresponding to the bonding / antibonding states.

2.4 Many-orbital cases

The formalism is easily generalized for many-orbital cases [15], such as sp-,d- and sd-valent systems. In these cases, Eq. (1) takes the form of

$$
U_{\text{bond}}^{i,j} = 2\sum_{\alpha,\beta} H_{i\alpha,j\beta}\Theta_{j\beta,i\alpha}, \qquad (26)
$$

where $\alpha(\beta)$ is an orbital on atom $i(j)$.

Taking the local quantization axis z along the bond direction, the bond matrix $H_{i\alpha,j\beta}$ will be simplified in compact blocks which arise from σ, π and δ bonds. A purely d bonded system, for example, has contribution from σ, π and δ bonds:

$$
U_{\text{bond}}^{i,j} = 2\left[H_{i,j}^{\sigma}\Theta_{j,i}^{\sigma} + 2H_{i,j}^{\pi}\Theta_{j,i}^{\pi} + 2H_{i,j}^{\delta}\Theta_{j,i}^{\delta}\right], \qquad (27)
$$

where

$$
\left.
\begin{aligned}
\Theta_{j,i}^{\sigma} &= \Theta_{j(3z^2-r^2),i(3z^2-r^2)}, \\
\Theta_{j,i}^{\pi} &= -\tfrac{1}{2}[\Theta_{jxz,ixz} + \Theta_{jyz,iyz}] \\
\Theta_{j,i}^{\delta} &= \tfrac{1}{2}[\Theta_{jxy,ixy} + \Theta_{j(x^2-y^2),i(x^2-y^2)}],
\end{aligned}
\right\} \qquad (28)
$$

and $H_{i,j}^{\nu}$ are appropriate Slater–Koster two-center bond integrals [16]. We note that the π bond order can be treated (similarly for δ bond order) choosing the generalized starting vector as

$$
|u_0^{\lambda,\pi}\rangle = \frac{1}{2}\left[\left(|ixz\rangle + e^{i\theta}|jxz\rangle\right)\mathbf{e}_1 + \left(|iyz\rangle + e^{i\theta}|jyz\rangle\right)\mathbf{e}_2\right], \qquad (29)
$$

where $\mathbf{e}_1$ and $\mathbf{e}_2$ are orthogonal unit vectors in an auxiliary two dimensional vector space. This treatment has the virtue that recursion coefficients are all invariant under the rotation of the local coordinate frame on the local z axis; thus we are allowed to take an arbitrary choice of local x(y) axis.

3. Approximate Model and Sum Rules

3.1 Sum rules

The theorem of many-atom expansion Eqs. (12) and (14) are only exact if summation is taken up to infinity. The first simplification to the bond order may be made by taking the partial sum of the many-atom series, namely

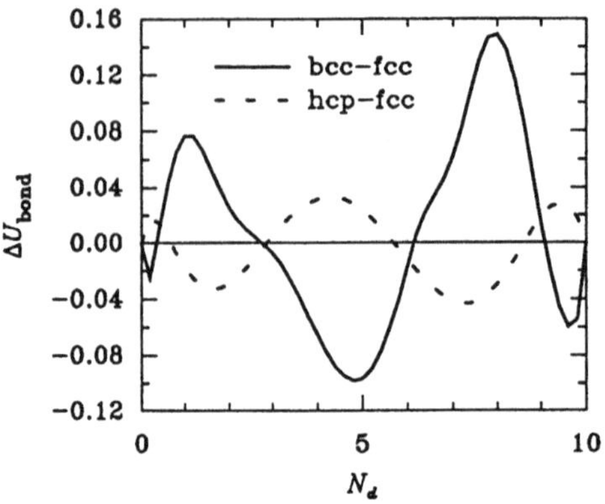

Figure 2. Bond energy differences for purely d bonded bcc, fcc and hcp lattices, calculated using the partial sum $\Theta^{[2,2]}_{i,j}$.

$$\Theta^{[l,m]}_{i,j} = -2\left[\sum_{n=0}^{l}\chi_{0n,n0}(E_{\rm F})\delta a_n + \sum_{n=1}^{m}\chi_{0(n-1),n0}(E_{\rm F})2\delta b_n\right]. \tag{30}$$

Figure 2 shows that a correct structural trend of transition metals can be reproduced in the bond energy differences calculated using the partial sum $\Theta^{[l,m]}_{i,j}$ to $l = m = 2$, where we took canonical d band parameters $(H^{\sigma}_{i,j}, H^{\pi}_{i,j}, H^{\delta}_{i,j}) = -(6,4,1)(2/5)\left(S/R_{ij}\right)^5$, where S is the Wigner-Seitz radius and R_{ij} is the bond length between atoms i and j. The convergence behavior of the bond order Eq. (30), which has to depend on the choice of a terminator, on increasing l and $m(= l, l+1)$ was found to be good [17, 18]. When terms up to $l = 7$ and $m = 8$ are taken into account, the series converges in good agreement with the exact value [19] calculated by tight binding bond model.

The first sum rule states that the interference terms must be related to the intersite Green's function as

$$\zeta_{m+1} = -\frac{1}{\pi}\int_{-\infty}^{\infty} E^m {\rm Im} G_{i,j}(E + i0)dE, \tag{31}$$

taking $a^0_0 = \mu_1 \equiv 0$ as our reference energy. We find

$$\zeta_2 = \delta a_0, \tag{32}$$

$$\zeta_3 = b^0_1 2\delta b_1, \tag{33}$$

$$\zeta_4 = 2(b^0_1)^2\delta a_0 + a^0_1 b^0_1 2\delta b_1 + (b^0_1)^2\delta a_1, \tag{34}$$

and so on, which are identical to Eq. (22)–(25). Applying this sum rule to the perturbatively derived many-atom expansion for the bond order [20, 21], we will be lead the same expression as Eq. (14) with Eq. (13).

The second sum rule can be found from the identity

$$(Z - E_i)G_{i,i}(Z) - \sum_{j(\neq i)} H_{i,j}G_{j,i}(Z) = 1. \tag{35}$$

It follows that the integrated number of electrons per spin at site i is given by

$$N_i(E_F) = -\frac{1}{\pi}\mathrm{Im}\left[\mathrm{P}\int_{-\infty}^{E_F}\sum_{j(\neq i)} H_{i,j}\frac{G_{j,i}(E+i0)}{E-E_i}dE\right] + \mathrm{Re}\left[1+\sum_{j(\neq i)} H_{i,j}G_{j,i}(E_i+i0)\right]\int_{-\infty}^{E_F}\delta(E-E_i)dE. \quad (36)$$

In order to avoid the δ-function discontinuity at $E = E_i$ when using an approximate intersite Green's function for a band continuum, we must have

$$1+\sum_{j(\neq i)} H_{i,j}\mathrm{Re}\left[G_{j,i}(E_i+i0)\right] = 0, \quad (37)$$

which is our second sum rule. It can also be found directly from Eq.(35) if we evaluate its real part at $Z = E_i$.

3.2 Symmetric fourth moment approximation

Taking into account correct even-order moments up to μ_4, the response functions become dependent on b_1^0 and b_2^0. In symmetric fourth moment approximation (SFMA), we set $a_0^0 = a_1^0 = a_2^0 = \cdots = 0$ and $b_1^0 \neq b_2^0 = \cdots = b_\infty$. The SFMA fulfills the second sum rule Eq. (37) for the case of s-orbitals on a lattice where all sites are equivalent. Since all δb_n's vanish by the band symmetry and also δa_n vanishes for $n \geq 2$ in a equivalent-site system, the bond order within the SFMA is given by

$$\Theta_{i,j}^{\mathrm{SFMA}} = -2\left[\chi_{00,00}(E_F)\zeta_2 + \chi_{01,10}(E_F)\frac{\zeta_4 - 2\mu_2\zeta_2}{\mu_2}\right]. \quad (38)$$

Denoting $\beta_1 = b_1^0/(2b_2^0)$ and $\varepsilon_\mathrm{F} = E_\mathrm{F}/(2b_2^0)$, [1] we obtain following analytic expressions for the response functions

$$\chi_{00,00}(E_\mathrm{F}) = \frac{\beta_1^2}{4b_2^0\pi}\left(\frac{1}{2}-\beta_1^2\right)V_1(\varepsilon_\mathrm{F}), \quad (39)$$

$$\chi_{00,10}(E_\mathrm{F}) = \frac{\beta_1}{4b_2^0\pi}U_0(\varepsilon_\mathrm{F}) + \frac{\beta_1^3}{4b_2^0\pi}\left[V_2(\varepsilon_\mathrm{F}) - 2\beta_1^2V_0(\varepsilon_\mathrm{F})\right], \quad (40)$$

$$\chi_{01,10}(E_\mathrm{F}) = \frac{\beta_1^2}{4b_2^0\pi}\left[V_3(\varepsilon_\mathrm{F}) - 2\beta_1^2V_1(\varepsilon_\mathrm{F})\right], \quad (41)$$

where

$$U_0(\varepsilon_\mathrm{F}) = \frac{1}{2i\beta_1^2\sqrt{\frac{1}{4}-\beta_1^2}}\left[I_0(\varepsilon_1;\varepsilon_\mathrm{F}) - I_0(-\varepsilon_1;\varepsilon_\mathrm{F})\right] \quad (42)$$

[1] One can choose either positive or negative b's. Our choice is the latter.

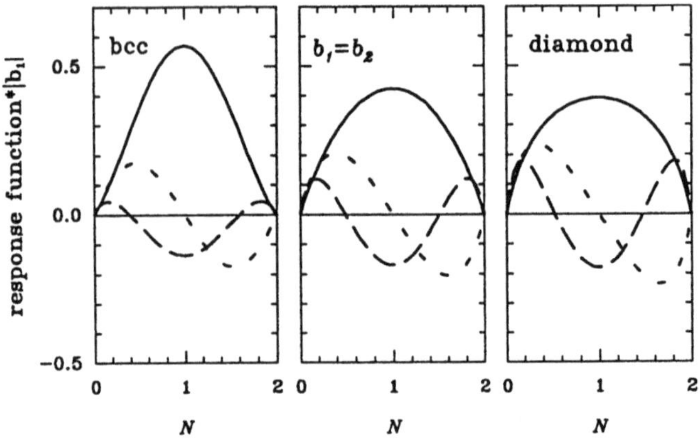

Figure 3. The response functions within the SFMA for three different values of b_2^0/b_1^0 (= 1.54, 0.87 for s bonded bcc and diamond lattices, respectively). $\chi_{00,00}|b_1^0|$, $\chi_{00,10}|b_1^0|$ and $\chi_{01,10}|b_1^0|$ are plotted as a function of the number of electrons per bond per spin in solid, dashed and long-dashed lines, respectively.

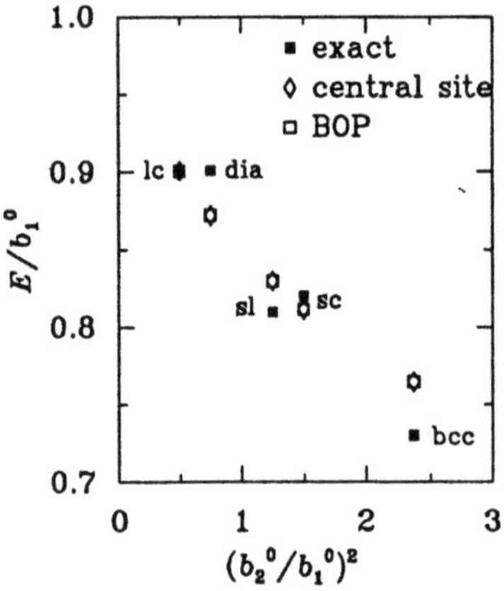

Figure 4. Dimensionless bond energy per atom versus dimensionless fourth moment within different approximations for a half-full band. Examples are taken from s bonded lattices: linear chain (lc), 2D square lattice (sl), diamond (dia), simple cubic (sc) and bcc.

and

$$V_n(\varepsilon_F) = -\frac{1}{4\beta_1^4(\frac{1}{4}-\beta_1^2)}\Bigg\{J_n(\varepsilon_1;\varepsilon_F)+J_n(-\varepsilon_1;\varepsilon_F) + \frac{i\sqrt{\frac{1}{4}-\beta_1^2}}{\beta_1^2}\left[I_n(\varepsilon_1;\varepsilon_F)-I_n(-\varepsilon_1;\varepsilon_F)\right]\Bigg\} \tag{43}$$

with $\varepsilon_1 = i\beta_1^2/\sqrt{\frac{1}{4}-\beta_1^2}$. The integrals I_n and J_n are defined by

$$I_n(z;\varepsilon_F) = \int_1^{\varepsilon_F} \frac{\varepsilon^n\sqrt{1-\varepsilon^2}}{\varepsilon - z}d\varepsilon \tag{44}$$

and

$$J_n(z;\varepsilon_F) = \int_1^{\varepsilon_F} \frac{\varepsilon^n\sqrt{1-\varepsilon^2}}{(\varepsilon - z)^2} d\varepsilon = \frac{\partial}{\partial z} I_n(z;\varepsilon_F). \tag{45}$$

Analytic expressions for these integrals is recursively obtained starting from the initial term given by

$$I_0(z;\varepsilon_F) = \sin\phi_F + z\phi_F - \sqrt{1-z^2}L(z;\phi_F), \tag{46}$$

with

$$L(z;\phi_F) = \ln\left[\frac{\sqrt{1-z}\cos(\phi_F/2) + \sqrt{1+z}\sin(\phi_F/2)}{\sqrt{1-z}\cos(\phi_F/2) - \sqrt{1+z}\sin(\phi_F/2)}\right], \tag{47}$$

where $\phi_F = \cos^{-1}(\varepsilon_F)$. The number of electrons per bond per spin is found

$$N = -(\beta_1^2/4\pi)U_0(\varepsilon_F). \tag{48}$$

A variation of SFMA response functions with the value of b_2^0/b_1^0 is displayed in three panels of Figure 3. The bond energy calculated with Eq. (38) gives well-convergent results [22] which are identical to those given by the central-site recursion method applied to s bonded lattices within SFMA [23] as shown in Figure 4. We note that analytic expressions given here for response functions and the number of electrons are strict if $0 < b_1^0/b_2^0 < \sqrt{2}$. For the case of $b_1^0/b_2^0 > \sqrt{2}$, we include additional contributions from a pair of delta functions which appear outside the band. Even in this case, expressions are given analytically [13].

4. Conclusion

We have derived new angularly-dependent many-atom bond order potentials for saturated and unsaturated bonds within the tight-binding Hückel approximation. In contrast with perturbatively derived many-atom potentials, the bond order potentials, which are firmly based on the exact many-atom expansion theorem in the present formalism, do not need to have any reference system. The angular functional forms of the bond order can be extracted from moments and interference terms about a bond. The approximate bond order potentials with truncated many-atom series are found to display as good a convergence as that given by approximate central-site recursion method if sum rules are satisfied. Such simplified potentials will appear powerful in modeling and predicting materials properties by atomistic simulations.

Acknowledgment

One of the authors (PG) wishes to thank the U.K. Science and Engineering Research Council for financial support.

References

1. *Atomistic Simulation of Materials*, edited by V. Vitek and D.J. Srolovitz, (Plenum, New York, 1989).
2. M.S. Daw and M.I. Baskes, Phys. Rev. B **29**, 6443 (1984).
3. M.S. Daw, Phys. Rev. B **39**, 7441 (1989).
4. J. Tersoff, Phys. Rev. B **37**, 6991 (1988); Phys. Rev. B **38**, 9902 (1988).
5. F. Stilinger and T. Weber, Phys. Rev. B **31**, 5262 (1985).
6. J.A. Moriarty, Phys. Rev. B **42**, 1609 (1990).
7. A.E. Carlsson, in *Solid State Physics*, edited by H. Ehrenreich and D. Turnbull (Academic, New York, 1990), Vol. 43, p. 1.
8. C.A. Coulson, Proc. Roy. Soc. (Lond.) A **169**, 413 (1939).
9. A.P. Sutton, M.W. Finnis, D.G. Pettifor, and Y. Ohta, J. Phys. C **21**, 35 (1988).
10. C. Lanczos, J. Res. Natl. Bur. Stand. **45**, 225 (1950).
11. R. Haydock, V. Heine, M.J. Kelly, J. Phys. C **5**, 2845 (1972).
12. R. Haydock, in *Solid State Physics*, edited by H. Ehrenreich and D. Turnbull (Academic, New York, 1980), Vol. 35, p. 215.
13. M. Aoki, P. Gumbsch and D.G. Pettifor, unpublished.
14. D.G. Pettifor and M. Aoki, in *Structural and Phase Stability of Alloys*, edited by J.L. Morán-López, F. Mejía-Lira and J.M. Sanchez, (Plenum, New York, 1992), p. 119.
15. D.G. Pettifor, in *Many-Atom Interactions in Solids*, edited by M. Manninen, R.M. Nieminena, and M.J. Puska, (Springer-Verlag, Now York, 1990), p. 64
16. J.C. Slater and G.F. Koster, Phys. Rev. **94**, 1498 (1954).
17. M. Aoki and D.G. Pettifor, to be published in *Proc. Int. Conf. Phys. Transition Metals*, Darmstadt, FRG, (1992).
18. M. Aoki, and D.G. Pettifor, to be published in *Proc. of CAMSE '92* (1992).
19. S. Glanville, A.T. Paxton and M.W. Finnis, J. Phys. F **18**, 693 (1988).
20. D.G. Pettifor, Phys. Rev. Lett. **63**, 2480 (1989).
21. D.G. Pettifor and M. Aoki, Phil. Trans. R. Soc. Lond. A **334**, 439 (1991).
22. M. Aoki, P. Gumbsch and D.G. Pettifor, to be published in *Proc. of CAMSE '92* (1992).
23. R.H. Brown and A.E. Carlsson, Phys. Rev. B **32**, 6125 (1985).

Augmented-Plane-Wave Force Calculations for Transition-Metal Systems

T. Oguchi

National Research Institute for Metals, Material Physics Division,
2-3-12 Nakameguro, Meguro-ku, Tokyo 153, Japan

Abstract. Important methodological aspects involved in an augmented-plane-wave (APW) based atomic-force method are presented with focusing on a frozen-augmentation approximation and properties of the APW basis functions. The accuracy and consistency of the method are examined in a frozen-phonon calculation of bulk Si. The method is applied for structural optimization and phonon calculations of transition-metal systems such as Cu surfaces, ternary Cu oxide and ferromagnetic Fe.

1. Introduction

Atomic-force methods within the local-density approximation to the density-functional theory have been successfully applied for structural optimization and phonon calculations of a wide variety of condensed matter systems, being incorporated recently into molecular dynamics simulations with use of the Car-Parrinello technique [1,2]. In most of these calculations and simulations, plane-wave basis functions and pseudopotentials were widely used because the atomic force which is a crucial ingredient in the methods can be easily and accurately computed by the Hellmann-Feynman theorem. The force acting on a particular nucleus, α, is simply given as the gradient of a classical electrostatic field at the nucleus site, $\mathbf{R}_\alpha$, induced by the other nucleus charges, eZ_β, and the self-consistent electron density, $n(\mathbf{r})$, as,

$$\mathbf{F}_\alpha^{HF} = -2Z_\alpha \frac{\delta}{\delta \mathbf{R}_\alpha} \left[\sum_{\beta \neq \alpha} \frac{Z_\beta}{|\mathbf{R}_\alpha - \mathbf{R}_\beta|} - \int d\mathbf{r} \frac{n(\mathbf{r})}{|\mathbf{r} - \mathbf{R}_\alpha|} \right]. \tag{1}$$

The Rydberg atomic units ($\hbar=1$, $m=1/2$ and $e^2=2$) are used throughout this paper.

It is well known, however, that there are serious limitations and difficulties when the methods are applied to certain systems which contain spatially localized electronic states such as $2p$ states of the first row elements and d states of the transition metals. A huge number of plane waves are needed to represent such electronic states accurately and the calculation becomes a computationally demanding task when applied to many-atom systems in which great interests in the structure and dynamics are involved. Furthermore, there is another difficulty in treating a spin-polarized system properly within the present plane-wave pseudopotential scheme. To make the atomic-force calculations possible for these elements as well as delocalized electronic systems, it is necessary to adopt localized orbitals as the basis set in an all-electron formalism, although development of ultra-soft pseudopotentials could be another possibility to overcome the difficulties [3].

Springer Series in Solid-State Sciences, Volume 114
Interatomic Potential and Structural Stability Eds.: K. Terakura and H. Akai

Once we use localized basis functions, a correction force term appears to be computed in addition to $\mathbf{F}_\alpha^{HF}$, as Pulay pointed out [4], and the total atomic force is then written as,

$$\mathbf{F}_\alpha = \mathbf{F}_\alpha^{HF} + \mathbf{F}_\alpha^{P}. \tag{2}$$

This Pulay correction force, $\mathbf{F}_\alpha^{P}$, comes out essentially from incompleteness of the basis set and insufficient self-consistent field (SCF) convergence, and practically from explicit dependence of the basis functions on the nuclear coordinates as,

$$\mathbf{F}_\alpha^{P} = -\sum_i^{occ.} \left[<\frac{\delta\psi_i}{\delta\mathbf{R}_\alpha} | H | \psi_i> + <\psi_i | H | \frac{\delta\psi_i}{\delta\mathbf{R}_\alpha}> \right], \tag{3}$$

where ψ_i is the i-th wave function and H is the one-electron hamiltonian in the Kohn-Sham equation. The Pulay force is often called the variational force since it vanishes when the Kohn-Sham equation is exactly solved in a variational sense. There are several choices among localized basis functions such as Gaussian-type, Slater-type, numerical atomic orbitals or augmented type. Here, we take an augmented-plane-wave (APW) set since it has very nice characteristic features which are useful to calculate the atomic force.

In the present paper, some methodological aspects of the APW atomic force method are briefly described with focusing on a frozen-augmentation approximation and special features of modified APW basis functions we adopt. In order to demonstrate the accuracy and internal consistency of the method, a frozen-phonon calculation is carried out for bulk Si, which has been extensively studied with the plane-wave pseudopotential approach. The APW force method is then applied to a surface-contraction problem of Cu, where very localized 3d states are involved as valence. Next example is structural optimization and phonon calculation for the famous high-T_c related Cu oxide, La_2CuO_4. Finally, a preliminary result is given as an extension of the method to a spin-polarized system. A frozen-phonon calculation of ferromagnetic Fe is presented.

2. Methodological Aspects

The formulation of the APW atomic force has been proposed by three groups, Soler and Williams [5,6], Yu, Singh and Krakauer [7], and Goedecker and Maschke [8]. In all these methods, a frozen-augmentation approximation is adopted, in which a change in the muffin-tin functions, which are defined as solutions of the radial Kohn-Sham equation inside the muffin-tin sphere, due to an infinitesimal atomic displacement is neglected. With this approximation, the derivative of APW, which appears in the Pulay force of Eq. (3), simply becomes the derivative of the structure factor times APW as,

$$<\frac{\delta}{\delta\mathbf{R}_\alpha} \phi_{\mathbf{k+G}}^{APW} | H | \psi_i> = i(\mathbf{k+G}) <\phi_{\mathbf{k+G}}^{APW} | H | \psi_i>, \tag{4}$$

since only the matching conditions on the sphere change due to the displacement.

The validity of the frozen-augmentation approximation has been intensively discussed and has been well proven for transition-metal elements [7].

Among the proposals, Yu *et al.* [7] and Goedecker and Maschke [8] used the conventional linear APW (LAPW) basis functions, while Soler and Williams [5,6] proposed slightly different LAPW. In this study, we choose Soler and Williams' modified LAPW basis set.

In the conventional LAPW formalism [9,10], the wave functions are expanded with plane waves in the interstitial region (I) and with spherical waves constructed from the radial functions at a particular energy, $R_{\alpha\ell}$, and their energy-derivative functions, $\dot{R}_{\alpha\ell}$, inside the muffin-tin sphere (S_α) as,

$$\phi^{APW}_{\mathbf{k+G}}(\mathbf{r}) = \tilde{\phi}^{\mathbf{k+G}}(\mathbf{r})\,\Theta(\mathbf{r}\in I) + \sum_\alpha \Theta(\mathbf{r}\in S_\alpha) \sum_{\ell m=0}^{\ell_{max}} \phi^{\mathbf{k+G}}_{\alpha\ell m}(\mathbf{r}-\mathbf{R}_\alpha), \tag{5}$$

$$\tilde{\phi}^{\mathbf{k+G}}(\mathbf{r}) = \Omega^{-1/2} e^{i(\mathbf{k+G})\cdot\mathbf{r}}, \tag{6}$$

$$\phi^{\mathbf{k+G}}_{\alpha\ell m}(\mathbf{r}) = \left[A^{\mathbf{k+G}}_{\alpha\ell m} R_{\alpha\ell m}(r) + B^{\mathbf{k+G}}_{\alpha\ell m} \dot{R}_{\alpha\ell m}(r)\right] Y_{\ell m}(\mathbf{r}hat), \tag{7}$$

with the Heaviside function, $\Theta(\mathbf{r})$, and the spherical harmonics, $Y_{\ell m}(\hat{\mathbf{r}})$. The coefficients of the linear combination in Eq. (7), $A^{\mathbf{k+G}}_{\alpha\ell m}$ and $B^{\mathbf{k+G}}_{\alpha\ell m}$, are determined by the continuity conditions to a plane wave with the wave vector **k**+**G** in magnitude and slope on the muffin-tin sphere. It should be noted that the spherical-wave components higher than the maximum angular momentum, ℓ_{max}, are not included inside the muffin-tin sphere. Due to this, ℓ_{max} is usually taken to be sufficiently large, typically 6 or 8 for the transition-metal elements, to reduce the discontinuity of the wave functions.

In the modified LAPW formalism [5,6], on the other hand, the plane-wave expansion is used in the whole space and the spherical waves up to ℓ_{max} are *augmented* inside the muffin-tin sphere as written as,

$$\phi^{APW}_{\mathbf{k+G}}(\mathbf{r}) = \tilde{\phi}^{\mathbf{k+G}}(\mathbf{r}) + \sum_\alpha \Theta(\mathbf{r}\in S_\alpha) \sum_{\ell m=0}^{\ell_{max}} \left[\phi^{\mathbf{k+G}}_{\alpha\ell m}(\mathbf{r}-\mathbf{R}_\alpha) - \tilde{\phi}^{\mathbf{k+G}}_{\alpha\ell m}(\mathbf{r}-\mathbf{R}_\alpha)\right]. \tag{8}$$

Here, $\tilde{\phi}^{\mathbf{k+G}}_{\alpha\ell m}$ is the (ℓm) component of the spherical wave expansion of the plane wave at the site α. Since the *unaugmented* spherical waves higher than ℓ_{max} are included inside the muffin-tin sphere in this case, rigorous continuity of the wave functions in amplitude and slope on the muffin-tin sphere is achieved, independent of ℓ_{max}. Therefore, one can choose ℓ_{max} as low as possible, typically 2 for the transition-metal elements.

Similarly, the charge density and potential functions, $n(\mathbf{r})$ and $v(\mathbf{r})$, are represented in a general form with the plane-wave expansion in the whole space and with the spherical-wave expansion inside the muffin-tin spheres as,

$$n(\mathbf{r}) = \tilde{n}(\mathbf{r}) + \sum_\alpha \Theta(\mathbf{r}\in S_\alpha) \sum_{\ell m=0}^{\ell'_{max}} \left[n_{\alpha\ell m}(\mathbf{r}-\mathbf{R}_\alpha) - \tilde{n}_{\alpha\ell m}(\mathbf{r}-\mathbf{R}_\alpha)\right], \tag{9}$$

$$v(\mathbf{r}) = \tilde{v}(\mathbf{r}) + \sum_\alpha \Theta(\mathbf{r}\in S_\alpha) \sum_{\ell m=0}^{\ell'_{max}} \left[v_{\alpha\ell m}(\mathbf{r}-\mathbf{R}_\alpha) - \tilde{v}_{\alpha\ell m}(\mathbf{r}-\mathbf{R}_\alpha)\right]. \tag{10}$$

With this representation, the potential function is exactly continuous on the muffin-tin sphere and thus the kinetic energy, $-\nabla^2\psi_i(\mathbf{r})$, also becomes continuous, resulting no muffin-tin surface integration term in the atomic force, which appears in case of the conventional LAPW formalism [7].

Consequently, the expression of the present LAPW atomic force is rather simple containing three terms as,

$$\mathbf{F}_\alpha = -\int d\mathbf{r}\ \nabla\tilde{v}_C(\mathbf{r})\ \tilde{n}_\alpha^{ps}(\mathbf{r})$$

$$-2\sum_i^{occ.} \mathrm{Re}\left[<\frac{\delta}{\delta\mathbf{R}_\alpha}\psi_i\,|H-\varepsilon_i|\,\psi_i>_\alpha - <\frac{\delta}{\delta\mathbf{R}_\alpha}\tilde{\psi}_i\,|\tilde{H}-\varepsilon_i|\,\tilde{\psi}_i>_\alpha\right]$$

$$= \tilde{\mathbf{F}}_\alpha + \mathbf{F}_\alpha^{MT} - \tilde{\mathbf{F}}_\alpha^{MT}, \tag{11}$$

where ε_i is the expectation value of the Kohn-Sham hamiltonian,

$$\varepsilon_i = \frac{<\psi_i\,|H|\,\psi_i>}{<\psi_i\,|\,\psi_i>}, \tag{12}$$

$< \cdots >_\alpha$ represents a volume integral inside the muffin-tin sphere α, $\tilde{v}_C(\mathbf{r})$ denotes the plane-wave representation part of the Coulomb potential involved in Eq. (10) and $\tilde{n}_\alpha^{ps}(\mathbf{r})$ is the so-called pseudo charge density associated with the atom α, which has been described in detail in Weinert's prescription of the full-potential method [11]. The derivative of the wave function can be easily computed in the frozen-augmentation approximation just as mentioned above. The second term in Eq (11) can be regarded as the Pulay force in the sense that it vanishes when the basis set is complete and SCF is perfectly met. The better convergences achieved on the basis set and SCF, the smaller magnitude of the second term in practice. However, the second term should not be excluded from the total force. It is found that the convergence of the total atomic force as functions of the number of APW and SCF iterations is much faster than each component in Eq. (11) as in the situation in the total energy. This indicates certain important cancellation existing among the three components.

The self consistency of the electronic system may be achieved by the steepest descent method. The plane-wave expansion coefficient of the i-th wave function, $C_{\mathbf{k+G},i}$, is updated every time step according to its residue,

$$\delta C_{\mathbf{k+G},i} = -<\phi_{\mathbf{k+G}}^{APW}\,|H-\varepsilon_i|\,\psi_i>\cdot\delta t, \tag{13}$$

followed by the Gram-Schmidt orthonormalization. We find that a preconditioning treatment proposed by Teter *et al.* [12] is very efficient to accelerate the SCF convergence in all the cases we have studied.

Let us describe briefly how to obtain the information on phonon from the atomic force calculations. The atomic force calculated for the atomic configuration near the equilibrium can be fitted to a polynomial form of the displacement of atom β in the j-direction, $\Delta R_{\beta j}$, as,

$$F_{\alpha i} = -\sum_{\beta j}\frac{\partial^2 E}{\partial R_{\alpha i}\,\partial R_{\beta j}}\,\Delta R_{\beta j} + O(\Delta R^2), \tag{14}$$

By diagonalizing the adiabatic dynamical matrix given as,

$$D_{\alpha i,\beta j} = (M_\alpha M_\beta)^{-1/2} \frac{\partial^2 E}{\partial R_{\alpha i}\, \partial R_{\beta j}}, \tag{15}$$

with the mass of atom β, M_β, phonon frequencies squared and amplitudes are obtained as the eigenvalues and eigenvectors, respectively.

Before applying the present APW method to transition-metal systems, its accuracy and internal consistency are examined by performing a frozen-phonon calculation of bulk Si. We consider the Γ optical phonon mode which corresponds to a breathing motion of the nearest neighbor bond along the <111> direction. By stretching and squeezing the bond by a small amount (typically 1% of the bond length), changes in the total energy and the atomic forces are independently calculated. The results are fitted to a polynomial of the displacement, u, like,

$$\Delta E = \frac{1}{2} k\, u^2 + 4\, k_{xyz} \left[\frac{u}{\sqrt{3}}\right]^3, \tag{16}$$

$$F = k\, u + \frac{4}{\sqrt{3}} k_{xyz}\, u^2. \tag{17}$$

Table I lists the calculated harmonic and anharmonic force constants, k and k_{xyz}, and derived phonon frequencies, along with those by the previous calculations and experiment [7,13]. The force constants obtained via the total energy and the atomic force agree satisfactorily well, implying very good internal consistency of the method. Table I also shows high accuracy which is comparable with that in the plane-wave pseudopotential and conventional LAPW calculations.

Table I. Comparisons of k, k_{xyz} and frequency of Γ-point optical phonon in Si calculated from total energy and atomic force with results of LAPW (ref. 7) and PWPP (ref. 13) calculations and experiment.

	k (Ry/a_B^2)	k_{xyz} (Ry/a_B^3)	frequency (THz)
present work			
total energy	0.5479	0.3726	15.22
force	0.5481	0.3875	15.22
LAPW			
total energy	0.5586	0.4026	15.37
force	0.5613	0.4030	15.40
PWPP			
total energy	0.5436	0.357	15.16
force	0.5422	0.355	15.14
experiment	0.5704	0.382	15.53

3. Applications

3.1 Surface Contraction of Cu

The first transition-metal system to which the present APW force method is applied is Cu surface. A preliminary result on this issue has been reported elsewhere [14]. The system consists of very localized valence 3*d* states and a vacuum region, providing a severe examination of the method.

It is widely known that all the metals show more or less surface contraction, i.e., a relaxation of the surface atomic layer toward the bulk side. The magnitude of the contraction depends strongly on the metal element and the surface index. In this study, we consider Cu low-index surfaces, (110), (001) and (111). A five-layer repeated slab geometry with a sufficient vacuum region is assumed to model these surface systems. The interatomic distance parallel to the surface is fixed to be the calculated equilibrium distance of bulk Cu, which is slightly smaller than the experimental value, and the interlayer spacings are optimized by the atomic forces.

Table II lists the results of the interlayer spacing relaxations, A_{1g} phonon frequencies and surface energies calculated for Cu (110), (001) and $(1\bar{1}1)$ slabs along with several experimental data. In general, the magnitudes and signs of the optimized contraction are in good agreement with experiments by ion scattering and electron diffraction techniques [15-17]. The rougher surface shows the larger contraction, which is consistent with a breaking-bond argument that a bond which looses more neighboring bonds becomes stronger, and also with an effective-medium argument that the surface atoms move inwards so as to recover the interstitial electron density reduced by cleavage [18].

We find that the lower-frequency phonon modes have large amplitude on the surface atoms, implying a sort of softening of the force constant regarding the vertical motion of the surface atoms, while the higher-frequency modes correspond

Table II. Relaxation of the interlayer spacings, A_{1g} phonon frequencies and surface energies calculated for Cu (110), (001) and $(1\bar{1}1)$ slabs. The bulk interlayer spacings for the (110), (001) and (111) faces are 1.25A and 1.76A and 2.15A, respectively. HEIS (high energy ion scattering), MEIS (medium energy ion scattering), LEED (low energy electron diffraction).

	Δd_{12} (%)	Δd_{23} (%)	ω_1 (THz)	ω_2 (THz)	$E_{surface}$ (erg/cm²)	Ref.
present work						
(110)	-8.3	+2.0	4.28	5.93	2310	
(001)	-4.0	-0.3	4.01	6.62	1390	
(111)	-1.0	-2.4	2.74	5.87	960	
experiments						
(110) HEIS	-5.3	+3.3				15
(110) MEIS	-7.5±1.5	+2.5±1.5				15
(110) LEED	-9.1±0.9	+2.1±0.3				15
(001) LEED	-1.2±1.7	+0.8±1.7				16
(111) LEED	-0.7±0.5					17
average face					1790	20

almost to bulk ones [19]. The softening might explain a reduction of the surface Debye temperature by 20-40% from the bulk value observed by an ion scattering experiment [15]. The calculated surface energies indicate strong surface-index dependence and have the same tendency as the magnitude of the surface contraction. The experimental value for average face [20] is just in between those theoretical values.

3.2 Structural Optimization of La_2CuO_4

The second example is structural optimization and phonon calculations of the famous high-T_c related Cu oxide, La_2CuO_4. We consider the simplest tetragonal unit-cell (K_2NiF_4 structure) for simplicity, although it is the high-temperature phase and transforms into orthorhombic or another complicated tetragonal phase at low temperatures. With the tetragonal K_2NiF_4 symmetry, the *z* coordinates of La and apical O ions are internal parameters to be determined. There are two A_{1g} phonon modes related to these two coordinates. The optimized internal parameters and A_{1g} phonon frequencies by an APW force calculation are in excellent agreement with neutron diffraction [21] and Raman experiment [22], as shown in Table III. The higher and lower frequency modes correspond almost to independent A_{1g} motions of apical O and La, respectively, and the coupling of the modes are very small. This may be due to large difference in the masses of La and O atoms. Those features are very much consistent with the results of LAPW total-energy frozen-phonon calculations by Cohen *et al.* [23].

Table III. Theoretical end experimental structural parameters and A_{1g} phonon frequencies of La_2CuO_4.

	z(La)	z(apical O)	A_{1g}	
	(c units)		(cm^{-1})	
present work	0.3634	0.1805	240	421
experiments	0.3619*	0.1827*	228**	429**

*: Ref. 21; **: Ref. 22.

3.3 Phonon Calculation of Ferromagnetic Fe

The formulation of the present APW atomic force is based on an all-electron scheme; core states are given in a self-consistent way as solutions of the Kohn-Sham equation with the spherically averaged potential inside the muffin-tin sphere. Therefore, an extension of the formulation to a spin-polarized system is rather straightforward within the local-spin-density formalism. Frozen-phonon calculations are performed for ferromagnetic Fe. We assume two types of unit-cell with the observed volume corresponding to H-point ([1,0,0]) and N-point ([½,½,0]) longitudinal modes. The result is listed in Table IV together with experiments [19] and agreement with experiments is again very much satisfactory.

Table IV. Calculated and experimental phonon frequencies at H and N points in ferromagnetic Fe.

mode	frequency (THz) theory (error)	experiment
H [1 0 0]	8.79 (+3%)	8.56
N [1/2 1/2 0] longitudinal	9.63 (+4%)	9.26

4. Concluding Remarks

Showing the results of several applications for structural optimization and phonon calculation, we conclude that the present APW force technique is very accurate, internally consistent and remarkably efficient for the systems containing the transition-metal as well as the first-row elements. The technique could be a promising approach for structural optimization and possibly dynamics of impurities, defects, surfaces, interfaces and other wide variety of large systems.

Acknowledgements

I have benefited greatly from close collaboration with Taizo Sasaki on the present issues. I thank Stefan Blügel, Peter Dederichs, Hiroshi Katayama-Yoshida and Kiyoyuki Terakura for stimulating discussions.

References

[1] R. Car and M. Parrinello, Phys. Rev. Lett. **55** 2471 (1985).
[2] For a review, see for example, T. Oguchi and T. Sasaki, Prog. Theore. Phys. Suppl. **103** 97 (1991).
[3] D. Vanderbilt, Phys. Rev. **B41** 7892 (1990).
[4] P. Pulay, Mod. Phys. **17** 197 (1969).
[5] J.M. Soler and A.R. Williams, Phys. Rev. **B40** 1560 (1989).
[6] J.M. Soler and A.R. Williams, Phys. Rev. **B42** 9728 (1990).
[7] R. Yu, D. Singh and H. Krakauer, Phys. Rev. **B43** 6411 (1991).
[8] S. Goedecker and K. Maschke, Phys. Rev. **B45** 1597 (1992).
[9] O.K. Andersen, Phys. Rev. **B12** 3060 (1975).
[10] D.D. Koelling and G.O. Arbman, J. Phys. F: Metal Phys. **5** 2041 (1975).
[11] M. Weinert, J. Math. Phys. **22** 2433 (1981).
[12] M.P. Teter, M.C. Payne and D.C. Allan, Phys. Rev. **B40** 12255 (1989).
[13] M.T. Yin and M.L. Cohen, Phys. Rev. **B26** 3259 (1982).
[14] T. Oguchi and T. Sasaki, Proc. CAMSE'92 (in Yokohama, 1992).
[15] For surface contraction of Cu (110) surface, see for example, M. Copel, T. Gustafsson, W.R. Graham and S.M. Yalisove, Phys. Rev. **B33** 8110 (1986).
[16] For surface contraction of Cu (001) surface, see for example, F. Jona, P. Jiang and P.M. Marcus, Surf. Sci. **192** 414 (1987).

[17] For surface contraction of Cu (111) surface, see for example, S.A. Lindgren, L. Wallden, J. Rundgren and P. Westrin, Phys. Rev. B**29** 576 (1984).
[18] K.W. Jacobsen, Comments Cond. Mat. Phys. **14** 129 (1988).
[19] See for example, *LANDOLT-BÖRNSTEIN New Series, Group-III* **13a** (Splinger-Verlag, Berlin, 1981) p.44.
[20] W.R. Tyson and W.A. Miller, Surf. Sci. **62** 267 (1977).
[21] J.M. Delgado, R.K. McMullan, G. Diaz de Delgado, B.J. Wuensch, P.J. Picone, H.P. Jenssen and D.R. Gabbe, Phys. Rev. B**37** 9343 (1988).
[22] W.H. Weber, C.R. Peters, B.M. Wanklyn, C. Chen and B.E. Watts, Phys. Rev. B**38** 917 (1988).
[23] R.E. Cohen, W.E. Pickett and H. Krakauer, Phys. Rev. Lett. **62** 831 (1989).

Structural Optimization with the KKR Method and Its Application to Interstitial Impurities in Mg

H. Akai

Department of Physics, Nara Medical University, Kashihara, Nara 634, Japan

Abstract. Structural optimization based on the Korringa-Kohn-Rostoker band structure calculation is applied to Mg–N and Mg–Li super-cell structures simulating Mg based interstitial impurity systems. It is demonstrated that the simultaneous optimization of both the electronic structure and the local lattice relaxation works fairly well in this scheme. Especially, the force calculation provides reliable information necessary for the structural optimization even in rather early stage of the iteration of the density functional self-consistency. Owing to the small super-cells used in the calculation, however, the resulting lattice relaxation of Mg atoms around the impurities turns out as big as 25~30%, which is considerably bigger than the experimental observation ~15% for Mg–N.

1. Introduction

Due to the development of experimental techniques using probe nuclei implanted into target materials, which enable us to detect very microscopic information of solids, it is now feasible to study local structures and dynamics associated with the formation of probe–host complexes. From theoretical point of view, this poses a new type of problems which calls for reliable description of both electronic structures and local ionic arrangements or dynamics. For metallic systems, not many studies have been done along this line so far. It is partly due to the fact that the pseudo-potential method which has been proven so powerful in semi-conductor physics, as is symbolized by the success of the celebrated Car–Parrinello method, has not been very much successful for metallic systems, especially for 2sp elements and transitional elements. Few years ago, we proposed the application of the full-potential Korringa–Kohn–Rostoker (FPKKR) method developed by Jülich group as one of the possible way to avoid the difficulties arising in the application of the Car–Parrinello-like procedure for transition metal systems [1].

The FPKKR method deals with the full-potential by expanding the shape-function representing the Wigner-Seitz cell into the spherical

Springer Series in Solid-State Sciences, Volume 114
Interatomic Potential and Structural Stability Eds.: K. Terakura and H. Akai

harmonics. Though clearly FPKKR would be one of the best ways to deal with the structural optimization, the shape-function for the deformed Voronoi cell still causes some practical problems to be solved before being incorporated into the molecular dynamics.

Certainly a less accurate but much easier way is the usual KKR method. Though there is no clear evidence that the spherical muffin-tin potential used in the KKR can reproduce a reasonable charge distribution when a lattice relaxation exists, it is rather plausible that, at least, the change in the band energy caused by the lattice relaxation could be mostly accounted for in the framework of KKR through rearrangement of assembly of the muffin-tin potentials. It may be especially the case when the relaxation around a probe atom is not extremely big and hence does not change the local lattice structure completely. In this case even the rigid muffin-tin potential model works for the calculation of the relaxation energy. For this reason the usual KKR method could be a good starting-point. The main purpose of the present paper is to discuss the possibility of the structural optimization by use of KKR and to demonstrate how the simultaneous relaxation in both the lattice and the electronic structure works in this scheme.

We apply the approach to the lattice relaxation of interstitial N and Li impurities implanted in hcp Mg. These systems are typical examples and also are quite important for the following reason: Many quadrupole coupling constants have been determined in the last few years for light impurities (Li, B, N, O, F) in metallic systems thanks to the development of a high precision quadrupole resonance technique [2,3]. One of the interesting applications is the measurements of electric field gradients (EFG) of light impurities in hcp metals; any substitutional or interstitial sites in hcp crystals have a non-vanishing EFG. Since EFG's in metallic systems are mostly determined by local electronic properties around the probe atom, they provide us with important pieces of information about the local environment of impurities. Conversely, if the EFG's are known from electronic structure calculations, the nuclear quadrupole moments — many of them, e.g. that of ^{12}N, are still unknown — can be determined from the observed nuclear quadrupole coupling constant. Since EFG is quite sensitive to both the local lattice relaxation and the details of the electronic structure, the calculation affording accurate determination of them could be essential.

Section 2 briefly outlines the theoretical framework and discusses some practical points which are important in performing the structural optimization with KKR. The results of the calculation obtained for the impurities in Mg are discussed in section 3. Some details of the force calculation in the present scheme are also discussed. Section 4 summarizes the conclusions.

2. Theoretical framework

We adopt a usual expression for the total energy in the framework of the local density functional approximation (LSD), making use of an effective potential $v(\boldsymbol{r})$, and further approximate the electrostatic energy as well as the exchange-correlation energy by those suitable for muffin-tin charges $n(r) = \{n_\uparrow(r), n_\downarrow(r)\}$, namely spherical inside the muffin-tin spheres and constant everywhere else [4]. The total energy E_{tot} per unit cell is then given by (in a.u.)

$$E_{\text{tot}}\{n\} = E_{\text{b}}\{n\} + E_{\text{dc}}\{n\},$$

where the band energy E_{b} and the remaining part E_{dc} are given respectively by

$$E_{\text{b}} = E_{\text{F}} \sum_i Z_i - \int_{-\infty}^{E_{\text{F}}} \text{d}EN(E)$$

and

$$\begin{aligned} E_{\text{dc}} = &-\sum_i \int \text{d}\boldsymbol{r} v_i(r) n_i(r) + \frac{1}{2} \sum_i \int \text{d}\boldsymbol{r} \int \text{d}\boldsymbol{r}' \frac{2n_i(r)n_i(r')}{|\boldsymbol{r}-\boldsymbol{r}'|} \\ &- \sum_i \int \text{d}\boldsymbol{r} \frac{2Z_i n_i(r)}{|\boldsymbol{r}-\boldsymbol{R}_i|} + \frac{1}{2} \sum_{ij} 2Q_i A_{ij} Q_j \\ &+ \sum_i \int \text{d}\boldsymbol{r} \epsilon_{\text{xc}}\left(n_i(r)\right) n_i(r) + \Omega_{\text{int}} \epsilon_{\text{xc}}(n_{\text{int}}) n_{\text{int}}. \end{aligned}$$

Here $N(E)$ is the number of states below E, i runs over the lattice sites in the unit cell, Z_i is the nuclear charge at the i-th site, Q_i is the total positive charge inside the i-th muffin-tin sphere (i.e., Z_i- number of electrons in the sphere), A_{ij} is the constant depending only on the lattice structure, and n_{int} and Ω_{int} are the interstitial charge (spin) density and the interstitial volume, respectively. The computational expression for A_{ij} is rather tedious, since the Ewald sum has to be exploited, but straightforward.

In the above, full charges (i.e., not necessarily a muffin-tin charge) are implicitly accounted for in E_{b} whereas those entering E_{dc} are all muffin-tin charges. Thus the total energy expression used here might not suit LSD in strict sense. However, in this way we can avoid further unnecessary approximations which might enter during the energy minimization process. This point is important because it guarantees a internal consistency which is essential for the accurate force calculation. In the present case minimizations of both E_{b} and E_{dc} are performed more or less exactly: First, in KKR, neither the partial wave expansion nor the plane wave expansion causes truncation errors in the band energy E_b for given atomic t-matrices. This is partly a consequence of the fact that KKR is a scattering method; it includes all the contribution of the incident partial waves even though only a finite number of scattering waves on each site are considered. In this respect KKR is much different from linearized methods like LMTO and ASW for which the truncation in the partial wave expansion causes some problems. The convergence with respect to the plane wave number, on the other hand, is guaranteed by the exact folding of the matrix spanned by infinite number of the reciprocal vectors down to the structural constant matrix of the angular momentum representation. Second, the variational principle for E_{dc} with respect to the charge inside each sphere $n(r)$ yields muffin-tin potentials $v(r)$ without any averaging procedure [4]. Thus we can safely equate the potential appearing in E_b and that in E_{dc} during the self-consistent determination of $n(r)$.

Correspondingly, the force acting on each nuclei is obtained by differentiating E_{tot} with respect to the i-th nuclear position $\boldsymbol{R}_i$. The resulting expression for the force on i-th nucleus turns out to be

$$\boldsymbol{F}_i = \int \mathrm{d}\boldsymbol{r} n_i(\boldsymbol{r}) \nabla v_i(r) - \sum_j Q_i \frac{\partial A_{ij}}{\partial \boldsymbol{R}_i} Q_j .$$

Here $n_i(\boldsymbol{r})$ is the non-muffin-tin charge arising as a result of the multiple scattering. The contribution of discontinuity in $v_i(r)$ arising at the muffin-tin surface should be correctly included in the integral.

Minimization of the total energy functional is performed by a self-consistent iteration about $n(r)$. In conventional band structure calculations, the energy eigenvalues and eigenvectors are obtained from a KKR matrix for each k vector and the information obtained is used to construct $n(r)$. From a computational point of view, however, it is not the best way; a much faster alternative is to construct $n(r)$ directly from the Green function. In this case $n(r)$ is obtained by integrating the Green function with respect to the energy parameter along a path in the physical sheet of the complex half plane up to the Fermi level E_F. Non-muffin-tin components of $n(r)$, which are used for the force calculation, are also obtained in the same way.

The advantage of the direct use of the Green function is that in this way we can avoid the time consuming process of finding zeros of the KKR determinant. A minor disadvantage may be that the information about the energy dispersion is completely missing in this scheme. However, it is clear that the energy dispersion has little relevance to the physical excitation spectrum. Anyhow we have entirely eliminated the process of obtaining the energy eigenvalues, reducing the operation needed in the calculation by factor proportional to the number of valence electrons (or atoms) in the unit cell. For more detailed description about the Green function method for constructing charges, see e.g. [1].

We perform the structural optimization inside the unit cell by use of a simple steepest descent. No dynamical process like the dynamical simulated annealing is exploited. We set the muffin-tin sphere radii somewhat smaller than the touching ones to avoid intersection between the spheres during relaxation. The main contribution to the force is the second term of the expression, namely the interatomic electrostatic force. Since the interatomic electrostatic force in the present scheme depends only on the integrated charges in each muffin-tin sphere, the convergence of the force is generally better than the convergence of the charge density and the total energy. This allows us to move the ionic position following a rather crude estimation of the force obtained from non-self-consistent charges during iterations towards full self-consistency. This point will be discussed later in section 3.

Practical details of the simultaneous relaxation of the ionic position and the electron charges could be much dependent on the nature of a system treated. As for Mg–N and Mg–Li systems discussed in the following sections, the procedure is rather straightforward; one positional relaxation (following the steepest decent) every tenth LSD iteration cycle guarantees smooth simultaneous convergence.

3. Application to light interstitials in Mg

There are several distinct interstitial sites available for hcp crystals. Experiments performed by Kitagawa *et al* [5] suggested that ^{12}N recoil implanted into Mg single crystal finally stopped at one of the well defined sites, likely one of the trigonal interstitial sites (see figure 1). Intuitively, however, the most favorable interstitial site seems to be the octahedral-like site (figure 1) which have the biggest interstitial volume. It is interesting to see if a lattice relaxation may reverse the naive intuition. We apply the structure optimization introduced in the preceding section to Mg–N and Mg–Li systems.

While the situation realized in the experiment is a dilute impurity limit, the present method can treat only periodic systems. By compromise, we use a super-cell which simulates impurity systems; figure 1 displays the super-cells used in the calculations. Impurity concentration of these super-cell structures is still high, far from the experimental situation; only the way to improve the situation is to use a larger unit

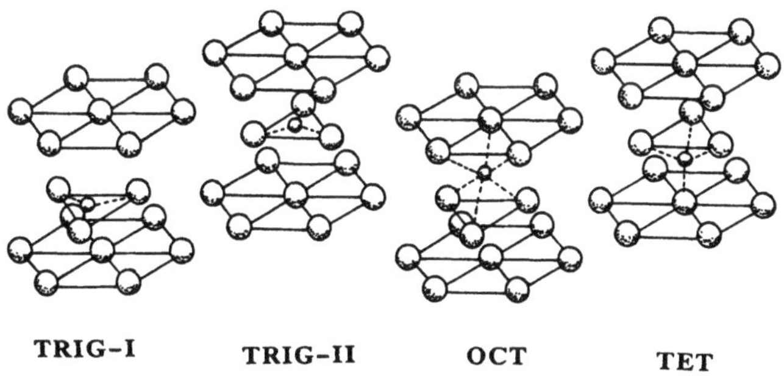

Figure 1. Unit cells used in the calculation. The larger and the smaller spheres represent the host Mg and the impurity atoms, respectively. Both TRIG-I and TRIG-II (Trig-I and Trig-II site in the text) have the trigonal symmetry, OCT (O site) has an octahedral-like and TET (T site) has a tetrahedral-like symmetry.

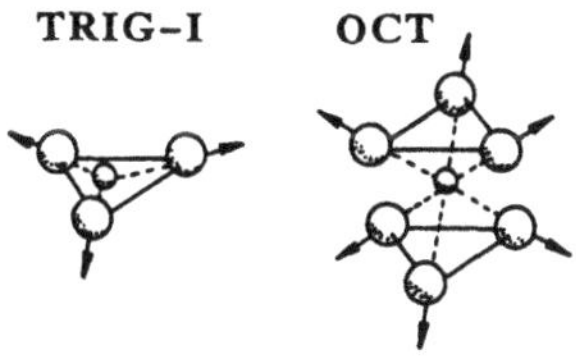

Figure 2. Relaxations around the impurity at TRIG-I (Trig-I site) and OCT (O site). The arrows indicate the direction of the relaxations.

cell, which, however, makes the computation increasingly difficult. For the time being, we think, the super-cells given in figure 1, admittedly not big enough, are still acceptable for the purpose of obtaining physical insight into the mechanism of various site occupations and the lattice relaxations.

We simulate the lattice relaxation by displacing the nearest neighbor host atoms, keeping the local symmetry around the impurity unchanged, as is shown in figure 2 by the arrows. Since it finally turns out that the energy differences among different site-occupations are rather large, the effect of the lattice relaxation is considered only for the two lowest energy impurity locations. Each of the muffin-tin radii is fixed throughout the calculation: $R_{\rm host} = 0.3464a$ and $R_{\rm impurity} = 0.2309a$, where a is the lattice constant, are used for all the cases. This makes direct comparisons of total energies among different atomic arrangements possible.

First we discuss the case of N implanted in Mg, only the case for which experimental information about the site occupation and the lattice

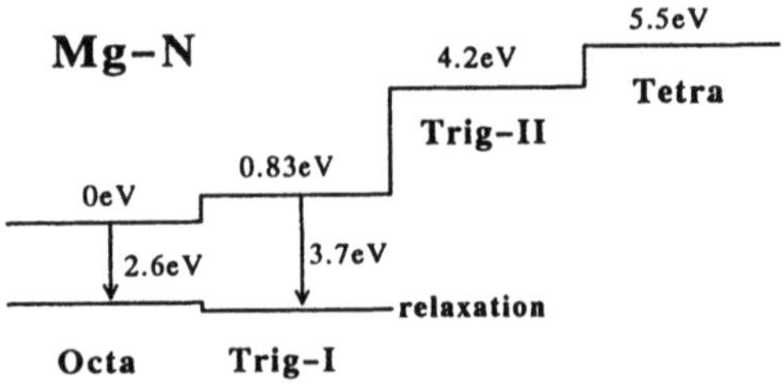

Figure 3. Energy diagram displaying the relative energies per unit cell (in eV) for various site occupation of N impurities in Mg. Also shown are the energies after a full lattice relaxation for the case of O and Trig-I sites.

relaxation is available. Figure 3 compared the calculated total energies for various N site occupations. As is naturally expected, the location of the lowest energy is the octahedral-like (O) site with the biggest interstitial volume. Though the tetrahedral-like (T) site also has a rather big interstitial volume, the energy for this location is the highest, 5.5 eV relative to the O site case. Among two trigonal sites, the center of the basal triangle just below the O site (Trig-I, see figure 1) gives a lower energy, being 0.83 eV relative to the O site.

The energy diagram displays also the energy gains due to the relaxation: they are 2.6 eV and 3.7 eV for the O site and Trig-I site, respectively. The relaxation for T and Trig-II sites were omitted. From the energy diagram we may conclude that the most stable impurity location among those considered here is the Trig-I site with relaxation. The energy difference between the O site and the Trig-I site is 0.27 eV.

Figure 4 shows the change in the total energy as a function of the displacement of nearest neighboring Mg atoms for the O and Trig-I sites. The minimum occurs at $\Delta/a \sim 0.06$ (O site) and ~ 0.17 (Trig-1 site), each corresponding to the local lattice expansion of about 10% and 30% respectively, where Δ is the displacement of the neighboring Mg atoms. Though seemingly the relaxations are much different between the two, the final equilibrium structures realized with these relaxations give almost the same impurity–host minimum distance.

The present result is quite consistent with the experimental indication that major part of N atoms (~85%) occupy one of the trigonal sites. As for the lattice relaxation around N impurities in Mg, Kitagawa *et al* [5] concluded from their experiments on orientation dependence of the dipolar broadening of the NMR spectrum against the external field that, the local lattice expansion around 15% was expected. Though the present result predicts much bigger lattice relaxation than the value suggested

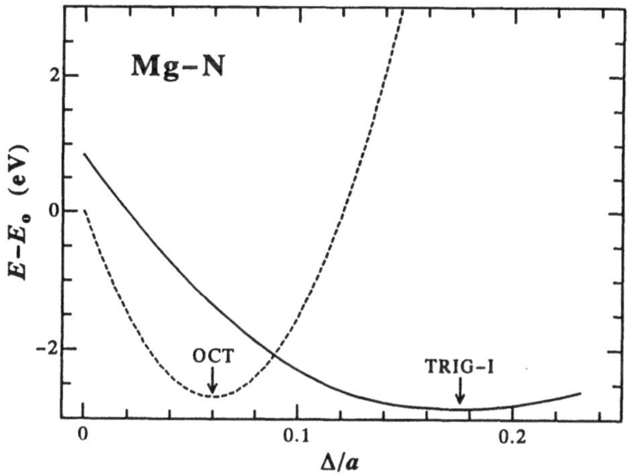

Figure 4. Change in the total energy of Mg–N as a function of the displacement (in the unit of the lattice constant) of nearest neighboring Mg atoms for the case of O (dashed line) and Trig-I sites (solid line). The arrows indicate the equilibrium lattice relaxation.

experimentally, we may say that the present result is reasonable in view that the calculation is performed for a rather small super-cell with a high impurity concentration.

A result of similar calculations for Li is shown in figure 5. Again the most stable site for the impurity is the Trig-I site with a lattice relaxation $\Delta/a \sim 0.15$, which is slightly smaller than the N case. From experimental point of view, Li in Mg is interesting since the nuclear quadrupole moment of Li has been determined through molecular beam experiments combined with quantum-chemical calculations using the configuration-interaction (CI) method and hence readily used to extract the information about electric field gradient (EFG) at the Li site. A recent NQR experiment [6] on ^{8}Li implanted in Mg then concludes that $|V_{zz}| = (3.81 \pm 0.39) \times 10^{18}$ (V/m^2)for Li. The present calculation gives $V_{zz} = +6.32 \times 10^{18}$ (V/m^2), which is somewhat larger than the experimental value. The unit cell used in the present calculation seems again too small. Another source of this discrepancy may be the muffin-tin potential model since the deformation of the single site potential contributes EFG considerably in some cases.

Finally. we examine the convergence of the force during the LSD iteration cycle in the present scheme and see if it can be used for estimation of the energy change along the relaxation path. Figure 6 shows the calculated force acting on one of Mg neighoring N and the deviation of the total energy from the converged one both against the

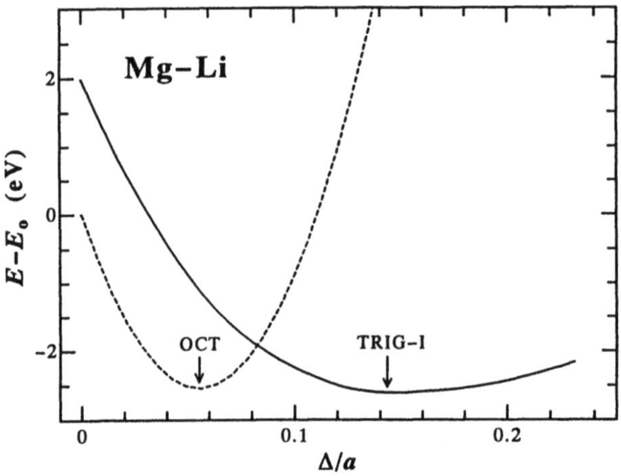

Figure 5. Change in the total energy of Mg–Li as a function of the displacement (in the unit of the lattice constant) of nearest neighboring Mg atoms for the case of O (dashed line) and Trig-I sites (solid line). The arrows indicate the equilibrium lattice relaxation.

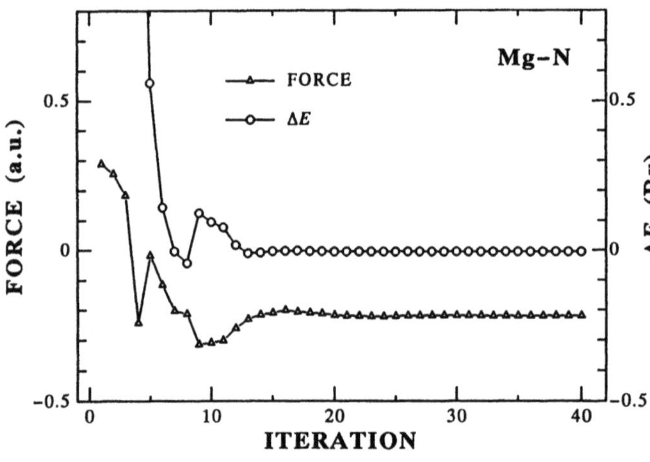

Figure 6. Calculated force (triangles) and deviation ΔE of the total energy from the fully converged one (circles) against the iteration step for Mg–N. The atomic positions are fixed to the unrelaxed ones.

iteration step. The atomic positions are fixed to the unrelaxed ones in this calculation. The force acting on one of the Mg atoms neighboring the N is plotted. From this figure, it is concluded that the convergence rate for the total energy is better than that of the force. This, however, generally does not mean that the force is less reliable than the total energy in each iteration step; contrary, we stress that the information about force is more reliable than the total energy in the early stage of the iteration. In other words, the convergence of the force is slower but seems better. The statement is based on the following observation: Though the fluctuation in the energy is not very big, say 0.1Ry, even in an early stage of the convergence, it is still comparable with a typical energy difference between two configurations, e.g., with and without a 10% lattice relaxation in the present case, making the comparison of the total energies meaningless. On the other hand, the fluctuation in the force, which amounts to around 50% in the similar stage of the convergence, can still predict the energy difference within 50% error with the correct sign, which is enough to see the stability of the lattice.

Similar, but may not be the same, observation has been reported by Oguchi [7] on modified LAPW force calculation; he states that the convergence of the force is faster than that of the total energy. For the present case the feature above originates from that the force is mostly determined by the interatomic electrostatic contribution, as is pointed out in the preceding section (the explanation, however, certainly depends on the way of partitioning of the total force into various contributions).

The muffin-tin radius of each component atom in the present calculation is chosen rather arbitrarily. Fortunately, both the band contribution and the electrostatic contribution to the force does not depends much on the muffin-tin radius unless we choose some extreme values (like half the atomic radius, etc.). The situation is understood by noticing that, first, the interstitial charge density is more or less constant which makes the electrostatic contribution independent of the muffin-tin radius and, second, the band contribution becomes conspicuous only in the vicinity of the nucleus. The second point is a consequence of the factor $\nabla v(r)$ entering in the expression of the force; an example is shown in figure 7 where the integrand appearing in the force expression $n_1(\boldsymbol{r})\nabla v(r)$, together with the $\ell = 1$ component (p-symmetry) of the charge density $n_1(\boldsymbol{r})$, for unrelaxed Mg–N is plotted as a function of r. Though n_1 has a big amplitude near the muffin-tin surface, corresponding force arises only near the origin. Moreover the large cancellation takes place due to its nodal structure; this is the very reason why the band contribution is so small.

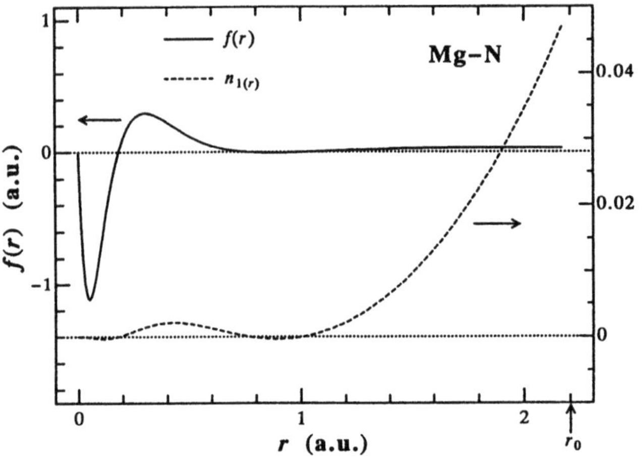

Figure 7. Charge density $n_1(\boldsymbol{r})$ with p-symmetry (dashed line) and the non-zero contribution $n_1(\boldsymbol{r})\nabla v(r)$ to the integral appearing in the force expression (solid line, see text), both as functions of r. Calculations for unrelaxed Mg–N.

4. Summary

The electronic structure of Mg–N and Mg–Li systems are calculated by use of the super-cell KKR method. Local structural optimization around the impurity is performed through a simultaneous relaxation with respect to both the atomic location and the electronic structure. The calculation predicts that, whereas the most stable site for the occupation of these impurities in the unrelaxed Mg lattice is the octahedral site, the local lattice relaxation lower the energy of the trigonal site case, finally making it the most stable location. The nearest neighbor lattice expansion thus occurs is about 30% for Mg–N and 26% for Mg–Li. These values are rather big but still seems consistent with experimental observation for N impurity implanted in Mg in view that the present calculation uses a small unit-cell to simulate the impurity systems.

The force calculation in the present approach is quite simple, introducing neither complications nor additional computation. Nice feature of the force calculation in the KKR scheme is that the convergence in the force seems better than the convergence in the total energy in early stage of the LSD iteration mainly due to dominance of interatomic electrostatic force, which makes the determination of the direction of the lattice relaxation without waiting full convergence possible.

Finally, we briefly comment on the applicability of the present method to other systems, especially transition metals. Our preliminary results obtained for Fe-X (X=B, C, N) show that the method basically works for those systems [8]. It is pointed out, however, that the choice of the muffin-tin radius causes some delicate problems for Fe–N; the magnetic state of Fe–N seems sensitive to this choice. In our opinion, it is a rather special (and certainly interesting) case, not implying a general drawback of the method. It is clear, however, that further elaboration, including the non-muffin-tin treatment and the use of bigger super-cells, is necessary for fully quantitative discussion. Improvement along this line together with the calculation of other light interstitials (B, F,...), as well as in other hcp host metals (Zn,...), are now in progress.

Acknowledgments

The author would like express his sincere thanks to Professor T. Minamisono, Professor T. Nojiri and Dr. T. Otsubo for stimulating discussions. The present work is partly supported by The Grant-in-Aid for Scientific Research of Priority Area, The Japanese Ministry of Education, Science and Culture.

References

[1] H. Akai, B. Drittler and P.H. Dederichs *Molecular Dynamics Simulations* F. Yonezawa(ed.), Springer, Berlin (1992) p177.

[2] Y. Someda, I. Minami, T. Otsubo, Y. Nakayama, A. Kitagawa, K. Matsuta, M. Fukuda, Y. Nojiri and T. Minamisono *Annual Report, Laboratory of Nuclear Studies, Osaka University* (1990) 68.

[3] T. Minamisono, T. Otsubo, I. Minami, S. Fukuda, A. Kitagawa, M. Fukuda, K. Matsuta, Y. Nojiri, S. Takeda H. Sagawa and K. Kitagawa *Phys. Rev. Lett.* **69** (1992) 2058.

[4] J.F. Janak *Phys. Rev.* **B9** (1974) 3985.

[5] A. Kitagawa, K. Matsuta, T. Nojiri and T. Minamisono *Hyperfine Int.* **60** (1990) 869.

[6] T. Otsubo, Y. Nakayama, I. Minami, M. Tanigaki, S. Fukuda, A. Kitagawa, M. Fukuda, K. Matsuta, Y. Nojiri, H. Akai and T. Minamisono *Hyperfine Int.* (1993) in press.

[7] T. Oguchi (1993) this volume.

[8] M. Takeda, H. Akai and J. Kanamori *Hyperfine Int.* (1993) in press.

Part II

Applications to Defects, Clusters and Surfaces

Conjugate-Gradient Total-Energy Minimization: Defects in Silicon

A. Oshiyama[1], *O. Sugino*[1], *and M. Saito*[2]

[1]Microelectronics Research Laboratories, NEC Corporation, 34 Miyukigaoka, Tsukuba 305, Japan

[2]NEC Scientific Information System Developement, Ltd., 34 Miyukigaoka, Tsukuba 305, Japan

We present the first-principles total-energy electronic-structure calculations for point defects in Si within the local density approximation. The highly efficient conjugate-gradient minimization technique is combined with the norm-conserving pseudo-potential momentum-space formalism, and enables us to perform large-scale calculations containing up to 216 atoms in a unit cell. The results provide rich information about the microscopic origins of the structural and electronic properties of defects in Si.

1 Introduction

Point defects and impurities play essential roles in structural and electronic properties of semiconductors[1]. Revealing microscopic origins of the phenomena related to the defects and the impurities is thus important not only scientific but also technological viewpoints. Total-energy electronic-structure calculations based on the first-principles of quantum mechanics have been contributing a lot in this field of defects in semiconductors. At present, most commonly used in the total-energy calculations is the supercell model in which each unit cell contains a point defect or an impurity and the unit cells are arranged periodically. Although the supercell model corresponds to the extremely high concentration of point defects in the materials, the computational technique in the band-structure calculations developed for the bulk materials is automatically transferable so that the energetics of the defects in semiconductors is described well in the supercell model. It is obvious, however, that we have to perform very large scale calculations in order to avoid artifacts due to interactions between different unit cells and to simulate point defects or impurities in an otherwise perfect crystal.

In density functional theory[2], we usually solve an effective one-electron equation (Kohn-Sham equation) which is Euler equation obtained from minimizing the total energy functional with respect to electron density. Solving the Kohn-Sham equation conventionally by diagonalizing the effective Hamiltonian matrix, however, requires a lot of computational efforts both in CPU time and in memories. An alternative is the iterative methods in which the total energy is directly minimized by improving the eigenvectors iteratively[3]. Among a variety of the iterative methods, the conjugate gradient method is a promising technique for minimization[4, 5].

In this paper, we report the total-energy electronic-structure calculations for defects in Si within the local density approximation (LDA) in density functional theory. The conjugate gradient method is used to minimize the total energy with respect to both electron and nucleus degrees of freedom. We start with the atomic and electronic structures of vacancy in Si, a paradigm of point defects in semiconductors, then discuss diffusion mechanisms of dopant impurities in Si.

Springer Series in Solid-State Sciences, Volume 114
Interatomic Potential and Structural Stability Eds.: K. Terakura and H. Akai

2 Calculation

There are four ingredients in the calculations. First, nuclei and core electrons are simulated by norm-conserving pseudopotentials[6]. The pseudopotential is constructed by the all-electron atomic (ionic) calculation, and is different for each angular momentum l (non-local pseudopotential). To save the computational time, we usually use Kleynman-Bylander (KB) approximation[7] which replaces the non-local and non-separable norm-conserving pseudopotential by a separable form. This approximation generally works well and provides the same result from the original norm-conserving pseudo-potential. Yet in some cases, e.g. Ga or As atoms in GaAs, the KB approximation induces ghost electron states [8] because the logarithmic derivatives of the pseudo-orbital at some energies are not always identical to the corresponding values from the all-electron orbital in KB approximation. In that case, we use real-space-partitioning procedure [9] in which the original pseudopotential is divided into several parts and then the KB procedure is performed for each part. This real-space-partitioning works well even in the case that the simple KB approximation dose not work.

Second, interactions among valence electrons are treated in the LDA in density functional theory. We use Ceperley-Alder form[10] for the exchange-correlation energy. The detailed description about the success and the limitation of LDA may be found elsewhere [11]. Third we use plane-wave basis-sets to express valence-electron wavefunctions and thus valence-electron density.

Fourth, we use supercell model in which each unit cell contains a point defect or/and an impurity and the unit cells are arranged periodically. Large-scale calculations are thus required to avoid artifacts due to interactions beyond boundaries of each cell and to simulate the defect in an otherwise perfect crystal. We use pre-conditioned conjugate gradient technique[5, 12] to minimize the total-energy with respect to the electron degrees of freedom at a given (ionic) geometry. Then the (ionic) geometry is relaxed to the lower total-energy geometry by the Hellmann-Feynman force and again using conjugate gradient technique. In that sense, we adopt the alternative minimization strategy [13] on the contrary to the simultaneous minimization[14]. In the optimized geometry, the calculated forces on each atom in the cell is less than typically 0.5 mRy/(atomic unit). Yet it may be fair to remind you that the size of the unit cell feasible at present is not enough for some purposes. For example, the vacancy in Si induces a deep level in the energy gap. In the 64-site supercell calculation, however, the deep level has the width of 0.6 eV due to the artificial interaction beyond the cell-boundaries. When we increase the size of the cell to 216 sites, the width decreases but is still 0.1 eV. Therefore we should be careful about the effect of this artificial band width on quantities which we like to calculate.

3 Vacancy in Silicon

The vacancy in Si is a paradigm of point defects in semiconductors: It induces deep levels in the energy gap, exhibits metastability rendering surrounding atoms relaxed, and plays a crucial role in atomic diffusion and aggregation[1]. In the ideal (i.e. unrelaxed) Si vacancy, a triply degenerate t_2 state appears in the middle of the energy gap, whereas an a_1 state resonates near the valence-band top. The t_2 state is partially occupied except for the doubly positive charge state, i.e., V^{2+}:a_1^2, V^{+}:$a_1^2t_2^1$, V^{0}:$a_1^2t_2^2$, V^{-}:$a_1^2t_2^3$, so that a symmetry-lowering lattice distortion takes place (Jahn-Teller effect). Many of the experimental data on the Si vacancy have been naively explained[1], or occasionally predicted[15, 16], by the one-electron theory including the Jahn-Teller distortion.

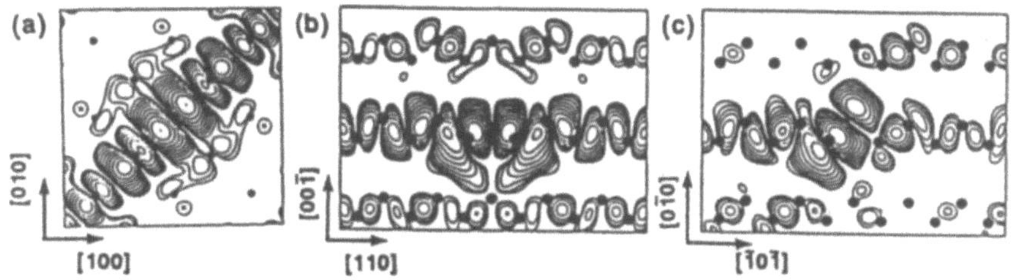

Figure 1: Contour plots of the squared wavefunctions of the b_1 state of the V^-; (a) in the (001) plane, (b) in the (110) plane and (c) in the (101) plane. The vacancy is situated at the center of each figure and the black dots denote the Si lattice sites. Three zigzag chains are seen in (b) and (c); the central chain corresponds to the chains A and B for (b) and (c), respectively. The maximum value of the plots is 2.8×10^{-2} Å^{-3} for (a) and 7.4×10^{-2} Å^{-3} for (b) and (c). The subsequent contour lines differ by a factor of 2.

Yet the success of the one-electron theory is not evident *a priori*. A many-electron effect is expected to be large in this system since the electrons in the deep levels are localized near the vacancy site. There is indeed an argument[17] that the many-electron ground state contains more than 20% (one-electron) excited configurations in V^-. Moreover, a recent model calculation shows that the experimentally observed C_{2v} symmetry of V^- can be explained only when the many-electron effect is taken into account[18]. The apparent contradiction between the success of the one-electron theory and the existence of the many-electron effect could be resolved if there exists an effective one-electron theory that includes essential features of the many-electron effect. The LDA is a good candidate for such an effective one-electron theory.

We are now in a position to present our LDA calculations[19]. The *symmetry-unrestricted* geometry optimization for V^- using 64-site supercell leads to the total-energy-minimized atomic geometry that has C_{2v} symmetry. In that geometry, the breathing distortion of the nearest-neighbor atoms is 0.23 Å and the Jahn-Teller distortion is several times smaller than that. This atomic configuration is satisfactorily reproduced by first distorting the neighboring 4 atoms inward by 0.20 Å and then moving two of the four atoms toward each other by 0.08 Å. Although this is essentially the same as the configuration speculated on the basis of a simple LCAO picture[1], the medium-range lattice relaxation is also important for the energetics of the vacancy: e.g., the Si atoms on zigzag chains in {110} planes are dislodged from the lattice site by 0.3, 0.1, 0.01 and 0.03 Å with increasing distance from the vacancy along the chain. The obtained level structure for V^- is again similar to the speculated one; there are three deep levels of a_1, b_1 and b_2 symmetries in the energy gap and the second level b_1 is half filled. An approximate level positions are 0.72 eV (a_1), 0.78 eV (b_1) and 0.80 eV (b_2) above the valence-band top. We have also optimized the geometry keeping the T_d symmetry, and found that its total energy is higher than that of the fully optimized one by 0.12 eV. This value can be compared with the measured bond reorientation energy of 0.07 eV [1]. All of these results are consistent with the experiments and thus we do not have to invoke the many-electron theory explicitly when we use the LDA.

We next investigate the deep-level wavefunction. Figure 1 is the squared wavefunction of the b_1 state of the V^- from the 216-site supercell calculation. Although the wavefunction is generally localized, it has very anisotropic extension along the zigzag chains in {110} planes: The amplitude of the wavefunction at the cell boundary along the chain

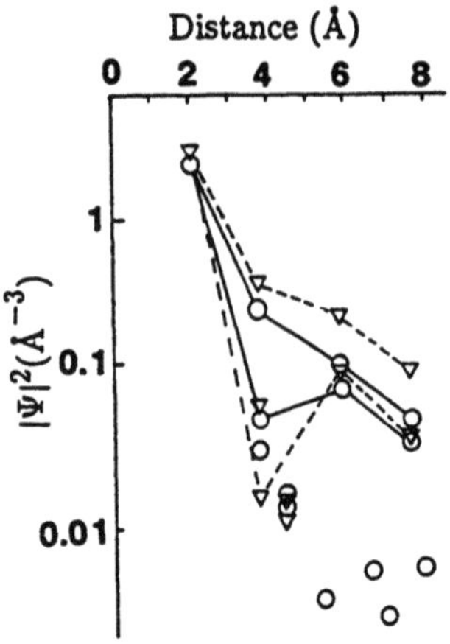

Figure 2: The amplitude of the b_1 wavefunction at nuclear sites plotted against the distance from the vacancy site. Calculated results are shown by circles and the ENDOR data of reference [20] are shown by triangles. The atomic sites on the zigzag chains are connected by solid lines (calculated results) and dashed lines (the ENDOR data); the chain A corresponds to the upper line and the chain B to the lower line.

is about 1/8 of the maximum value. Figure 1 also shows that there are two kinds of chains. One is that which extends in the [110] direction (Figure 1b), chain A hereafter, and the other is two equivalent chains extending in the [101] and [011] directions (Figure 1c), chain B hereafter. The amplitudes along the chains A and B are approximately the same. Yet there is an important difference for the second neighbor site of the vacancy, as is evident from the contour map in Figure 1.

Since an electron accommodated in the b_1 state gives rise to the spin density, we can compare the calculated wavefunction with the electron-nuclear-double-resonance (ENDOR) measurements. For that purpose, the calculation of the hyperfine tensor at each atomic site and then its assignment to the experimental values are required. The hyperfine tensor at site τ, A^{τ}_{ij}, can be expressed as a sum of the Fermi contact interaction term $a^{\tau}\delta_{ij}$, where $a^{\tau} = (8\pi/3)gg_N\mu\mu_N \mid \Psi(0) \mid$, and the dipole-dipole interaction term $B^{\tau}_{ij} = gg_N\mu\mu_N\langle\Psi \mid 3x_ix_j/r^5 - \delta_{ij}/r^3 \mid \Psi\rangle$. Here $g(g_N)$ is the electron (nuclear) g factor and $\mu(\mu_N)$ the electronic (nuclear) magneton. What we have obtained from the supercell calculations is the pseudo-wavefunction $\Psi^{ps}(\mathbf{r})$ for the b_1 state which is different from the all-electron wavefunction $\Psi(\mathbf{r})$ in the core region. Therefore we first expand $\Psi^{ps}(\mathbf{r})$ in terms of atomic pseudo-orbitals $\phi^{ps}_{lm}(r)$ as $\Psi^{ps}(\mathbf{r}) = \sum_{lm} d^{\tau}_{lm}\phi^{ps}_{lm}(\mid \mathbf{r}-\mathbf{R}_{\tau} \mid)Y_{lm}(\mathbf{r}-\mathbf{R}_{\tau})$, and second calculate atomic all-electron orbitals $\phi_{lm}(r)$ using logarithmic derivatives of $\phi^{ps}_{lm}(r)$. Then we obtain the all-electron wavefunction as

$$\Psi(\mathbf{r}) = \sum_{lm} d^{\tau}_{lm}\phi_{lm}(\mid \mathbf{r}-\mathbf{R}_{\tau} \mid)Y_{lm}(\mathbf{r}-\mathbf{R}_{\tau}).$$

Using this wavefunction $\Psi(\mathbf{r})$, we have calculated the hyperfine tensor at each nuclear site, and assigned it to the ENDOR data[20]. The results are summarized in Figure 2 which shows the wavefunction amplitude at each nuclear site determined from the calculation and the experiment. As is stated above, the amplitude along the zigzag chains are much larger than the value at another sites off the chains, and the values on the chains A and B are different at the second-neighbor sites. The agreement between the ENDOR data and the LDA calculations is excellent. This success, along with the fact that the LDA calculation provides the correct C_{2v} symmetry for the stable geometry, indicates the validity of the effective one-electron theory. It is still an open question whether the many-electron effect improve the agreement [19].

4 Impurity Diffusion in Silicon

Atomic diffusion, in particular the diffusion of dopant impurities, in Si is a technologically important and scientifically interesting issue[21]. The central problem not revealed yet is the microscopic mechanism of the diffusion. Intrinsic point defects, i.e. the vacancy and the self-interstitial atom (an additional Si atom in the interstitial region), mediate the diffusion of the dopant impurity. But which point defect is dominant in the diffusion of each impurity? The local density calculations in the past[22, 23] have revealed that the activation energies for the vacancy-mediated and the interstitial-mediated diffusions are comparable with each other in many cases. (The activation energy is a sum of the formation energy of the relevant defect and the migration energy of the diffusing species.) Thus, direct calculations of the diffusion coefficient which includes both enthalpy and entropy terms is required to determine the dominant mechanism. Yet we like to have another stance[24]: External pressure may resolve the mechanism since the two intrinsic point defects respond to the pressure in the opposite way, as is explained below.

Figure 3 is the calculated total energy along the migration path in the vacancy-mediated self diffusion. The diffusing atom moves to the neighboring vacant site by crossing the energy barrier at the middle point. The calculated migration energy is 0.2 eV in agreement with the experiment[1]. When we reduce the lattice constant, this migration energy decreases. Under the pressure of 60 kbar, the migration energy disappears. This is a consequence of the fact that Si atoms energetically favor more bonds than four under pressure: With increasing pressure, Si crystal exhibits a phase transition from the diamond structure to the β tin structure in which each Si has almost six nearest neighbors. In the

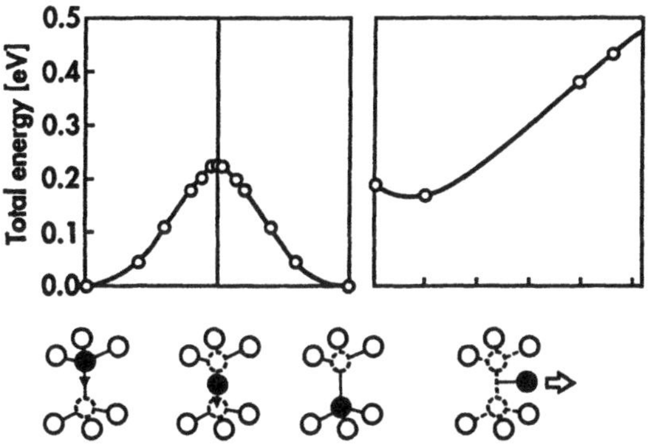

Figure 3: Calculated total energy variation along the migration path of the vacancy-mediated self-diffusion (diffusion of a Si atom in Si). The diffusing atom (•) moves to the neighboring vacant site (dotted circle) by crossing the energy barrier at the middle point (left panel). When the diffusing atom moves perpendicular to the bond axis, the total energy increases (right panel).

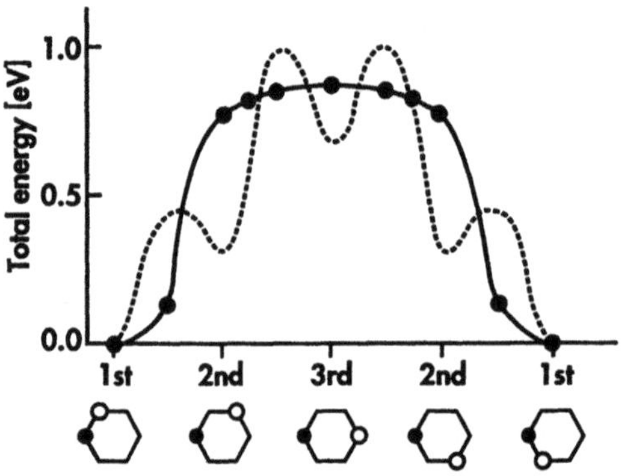

Figure 4: Calculated total energy in the migration process of neutral As-V pair. The vacancy (o) migrates around the As impurity (•) from one nearest neighbor site to the other nearest neighbor site. The filled circles are the calculated values and the solid line is a guide for eyes. The dashed line is the total energy variation imagined before.

Table 1: The activation energy for the vacancy mechanism at zero pressure (Q) and its variation at $P = 60kbar$ (ΔQ). The migration energy (E_m) of the impurity-vacancy pair is also shown. The experimental data are in parentheses. All quantities are in eV and a neutral charge state is assumed.

Species	Q	ΔQ	E_m
P	4.2 (3.5 ~ 3.7[a])	-0.9	0.8 (0.94[c])
As	3.9 (4.1 ~ 4.3[a])	-0.6 (-1.4[b])	0.8 (1.07[c])
Sb	3.9 (3.9 ~ 4.1[a])	-0.7	0.9 (1.28[c])
Si	4.2 (4.0 ~ 5.1[a])	-1.1	0.2 (0.45[d])

[a] Selected data from Reference [21].
[b] Compiled value from Reference [26].
[c] References [27, 28].
[d] Reference [29].

migration process shown in Figure 3, at first the Si atom has three neighbors, then six neighbors at the saddle point, and again three neighbors. This process has energy barrier without pressure, but under pressure the saddle point becomes energetically favorable.

We observe similar behaviors even without pressure. Figure 4 is the calculated potential energy surface in the diffusion of a pair of As impurity and the vacancy. (In the vacancy-mediated impurity diffusion, the diffusing species are the impurity-vacancy pair[25].) Naive expectation for the total-energy variation is the existence of several saddle points at the bond center position as is shown in the dashed line in Figure 4. Yet the calculated total-energy surface is smooth and exhibits only one saddle point. This is a consequence from the local stress caused by the As atom: Atomic radius of As is larger than that of Si so that the Si atoms around the As impurity feels the local stress. This stress renders the six-coordinated configuration (i.e. the bond center configuration) energetically favorable. The local stress caused by the impurity plays a similar role to the external pressure.

Decrease of the formation energy of the vacancy is more significant than the migration-energy change. For V^0, the formation energy changes from 4.0 eV to 3.0 eV under 60 kbar pressure (Table 1). This decrease is related to the lattice relaxation around the

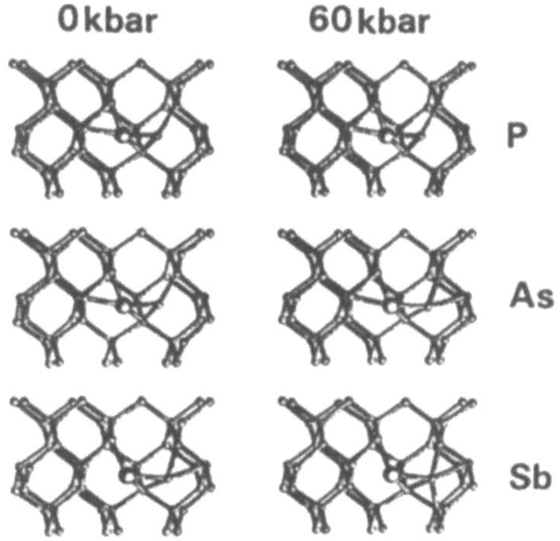

Figure 5: The most stable interstitial structure of impurity atoms (P, As and Sb) at $P = 0$ and $60kbar$. The charge states are neutral.

Table 2: The formation (E_f) and migration energy (E_m), and the activation energy ($Q = E_f + E_m$) for the interstitial at zero pressure and their energy increase at $P = 60kbar$. All quantities are in eV and a neutral charge state is assumed.

Species	E_f	Q	ΔE_f	ΔQ
P	3.5		0.5	
As	3.4	4.3	0.7	$\geq$0.4
Sb	3.9		0.1	

vacancy. The nearest neighbor Si atoms relax toward the vacancy by about 0.3 Å under zero pressure. As a result of this inward breathing distortion, the surrounding Si-Si bonds are stretched: the amounts of the stretching are 0.16 Å, 0.09 Å, 0.06 Å, respectively, with increasing distance from the vacancy. This stretching results in the tensile stress around the vacancy. By adding the external pressure this tensile stress is partially relaxed. This is the reason of the significant decrease of the formation energy. The activation energy (the formation energy plus the migration energy) for the vacancy-mediated diffusion is thus reduced significantly under pressure (Table1).

The interstitial atom responds to the pressure in the opposite way. Figure 5 is the calculated stable geometries for the interstitial impurity atoms. An interesting feature in these most stable geometries is the lack of symmetry. This is reasonable since the interstitial atom at high symmetry site generally induces degenerate deep levels partially occupied, and unstable against symmetry-lowering distortion. Around the interstitial atom, the Si-Si bonds are contracted by 0.05 Å$\sim$ 0.1 Å so that there is compressive stress around the interstitial atom. This stress is enhanced by adding the external pressure. In fact, the formation energy of the interstitial impurity atom increases under external pressure (Table 2), on the contrary to the vacancy case. Figure 6 is a migration path for the As interstitial impurity. The As migrates along the low symmetry path. The calculated migration energy is 0.7 eV. Although we have not calculated the migration energy for P and Sb, we expect the same order of magnitude for these impurity atoms.

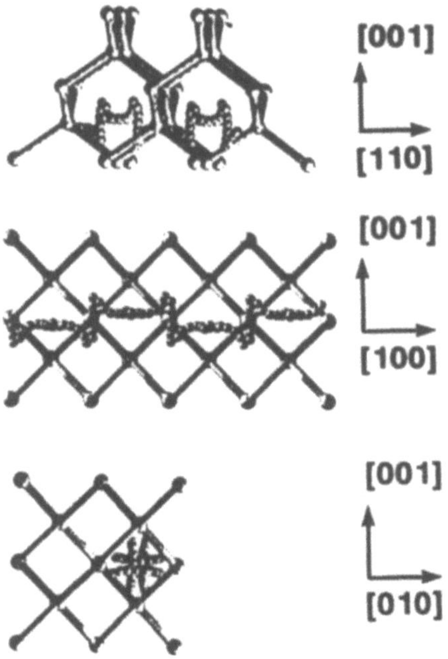

Figure 6: A favorable migration path of the interstitial As impurity seen from three different directions. The relaxation of the surrounding Si atoms is fully taken into account in the calculation, but not shown in this figure.

The variation of the activation energy under pressure comes mainly from the variation of the formation energy, and the direction is contrary to the case of the vacancy-mediated diffusion.

We are now in a position to compare the result with the experiment under pressure. The only data available is obtained for As impurity[26]. According to the experiment, the activation energy decreases with increasing pressure and $\Delta Q/\Delta P = -2.3 \pm 1.8 \times 10^{-2}$ eV/kbar. The calculated value for the vacancy-mediated As diffusion is $\Delta Q/\Delta P = -1.0 \times 10^{-2}$ eV/kbar, and the value for the interstitial mechanism is positive. The vacancy mechanism alone explains the observed decrease of the activation energy under pressure. The observed activation volume, $\Delta V = -9 \pm 4\AA^3$, is also consistent with the vacancy mechanism. It is true that the vacancy mechanism gives positive ΔV unless the lattice relaxation is considered. However, the neighboring atoms around the vacancy are relaxed toward the vacancy in the present case, and the calculated ΔV is equal to $-5\AA^3$.

What we like to propose here is that the calculated pressure dependence combined with the experiments under pressure may reveal the dominant diffusion mechanism for each impurity. Further, in current semiconductor devices, there are a lot of mismatched interfaces producing local stresses. Impurity diffusion in that system is expected to be different from the diffusion in bulk materials.

5 Conclusion

We have performed the first-principles total-energy electronic-structure calculations for defects in Si. The highly efficient conjugate-gradient technique is combined with the LDA pseudo-potential momentum-space formalism, and enables the large-scale calculations. The results have revealed the validity of the effective one-electron theory for the electron states induced by the vacancy, and microscopic mechanisms of the impurity diffusion under external pressure.

References

[1] See, for a review, *Deep Centers in Semiconductors* edited by S. T. Pantelides (Gordon and Breach, NY 1986)

[2] P. Hohenberg and W. Kohn, Phys. Rev. **136**, B864 (1964); W. Kohn and L. J. Sham, Phys. Rev. **140**, A1133 (1965).

[3] See, for example, L. A. Hageman and D. M. Young, *Applied Iterative Method* (Academic, 1981).

[4] I. Štich, R. Car, M. Parrinello, and S. Baroni, Phys. Rev. B**39**, 4997 (1989).

[5] M. P. Teter, M. C. Payne, and D. C. Allen, Phys. Rev. B**40**, 12255 (1989).

[6] D. R. Hamann, M. Schlüter, and C. Chiang, Phys. Rev. Lett. **43**, 1494 (1979); G. B. Bachelet, D. R. Hamann, and M. Schlüter, Phys. Rev. B **26**, 4199 (1982).

[7] L. Kleinman and D. M. Bylander, Phys. Rev. Lett. **48**, 1425 (1982).

[8] X. Gonze, P. Käckell, and M. Scheffler, Phys. Rev. B**41**, 12264 (1990).

[9] M. Saito, O. Sugino and A. Oshiyama, Phys. Rev. B**46**, 2606 (1992).

[10] D. M. Ceperley and B. J. Alder, Phys. Rev. Lett. **45**, 566 (1980).

[11] For example, *Theory of the Inhomogeneous Electron Gas* edited by S. Lundqvist and N. H. March (Plenum NY, 1983).

[12] D. M. Bylander, L. Kleinman, and S. Lee, Phys. Rev. B**42**, 1394 (1990).

[13] O. Sugino and A. Oshiyama, *Proc. 16th Int. Conf. Defects in Semiconductors* edited by G. Davis, G. G. Deleo, and M. Stavola (Trans Tech 1992) p469.

[14] R. Car and M. Parrinello, Phys. Rev. Lett. **55**, 2471 (1985).

[15] G. A. Baraff, E. O. Kane, and M. Schlüter, Phys. Rev. Lett. **43**, 956 (1979); Phys. Rev. B**21**, 5662 (1980).

[16] G. D. Watkins and J. R. Troxell, Phys. Rev. Lett. **44**, 593 (1980).

[17] M. Lannoo, Phys. Rev. B**28**, 2403 (1983).

[18] F. G. Anderson and F. S. Ham, *Proc. 16th Int. Conf. Defects in Semiconductors* edited by G. Davis, G. G. Deleo, and M. Stavola (Trans Tech 1992) p475.

[19] O. Sugino and A. Oshiyama, Phys. Rev. Lett. **68**, 1858 (1992).

[20] M. Sprenger, S. H. Muller, and C. A. J. Ammerlaan, Physica (Amsterdam) **116B**, 224 (1983); M. Sprenger, S. H. Muller, E. G. Sieverts, and C. A. J. Ammerllan, Phys. Rev. B**35**, 1566 (1987).

[21] For a review, P. Fahey, P. B. Griffin, and J. D. Plummer, Rev. Mod. Phys. **61**, 289 (1989).

[22] R. Car, P. J. Kelly, A. Oshiyama, and S. T. Pantelides, Phys. Rev. Lett. **52**, 1814 (1984); ibid., **54**, 360 (1985).

[23] C. S. Nichols, C. G. Van de Walle, and S. T. Pantelides, Phys. Rev. Lett. **62**, 1049 (1989); Phys. Rev. **40**, 5484 (1989).

[24] O. Sugino and A. Oshiyama, Phys. Rev. **B46**, No. 19, in press (1992).

[25] S. M. Hu, in *Atomic Diffusion in Semiconductors*, edited by D. Shaw (Plenum, NY, 1973) p217.

[26] E. Nygren, M. J. Aziz, D. Turnbull, J. M. Poate, D. C. Jacobson, and R. Hull, Appl. Phys. Lett. **47**, 105 (1985).

[27] G. D. Watkins and J. W. Corbett, Phys. Rev. **134**, 1359 (1964).

[28] M. Hirata, M. Hirata, and H. Saito, J. Phys. Soc. Jpn. **27**, 405 (1969).

[29] G. D. Watkins, J. R. Troxell, and A. P. Chatterjee, in *Defects and Radiation Effects in Semiconductors 1978* IOP Conf. Proc. No 46 (IOP, London, 1979) p16.

Structural Trends in Clusters and Disordered Systems

R.O. Jones

Institut für Festkörperforschung, Forschungszentrum Jülich, W-5170 Jülich, Germany

Abstract. The combined molecular dynamics/ density functional (MD/DF) method has been used to calculate the structures of small clusters and structural properties of amorphous and liquid materials. The focus in the present lecture is on two particular aspects of the results: (a) the relationship between the structures of clusters and extended systems of the same element, with selenium and phosphorus as examples. (b) The relationship between the structures of different elements of the same main group. Examples are taken from groups IIIa (Al, Ga), Va (P, As) and VIa (S, Se).

1. Introduction

In principle, molecular and condensed matter structures can be determined if one can calculate the total energy of the system of electrons and ions E for a given set of ionic coordinates R_I. One has "only" to repeat this calculation for all possible geometries, and identify the ground state structure as that with the lowest energy. It is natural to ask whether standard methods of molecular physics can be used, since the total energy of an interacting system of electrons and ions can be determined from the exact wave function Ψ. Such calculations can be carried out exactly for very few systems, however, and approximations for Ψ are generally unavoidable. The standard Hartree-Fock approximation, where Ψ is represented as a single (Slater) determinant of one-electron functions, can be improved systematically by incorporating additional determinants ("configuration interaction", CI). This leads, in principle, to an exact wave function and energy, but also to an explosive increase in the numerical effort required as the number of electrons increases.

These remarks apply to the energy calculation for a *single* geometry. In a cluster with N atoms, with $3N-6$ independent interatomic degrees of freedom, the number of local minima in the energy surface can be very large indeed. For clusters of atoms with a pairwise Lennard-Jones interaction, Hoare and McInnes [1] found that the number of minima increases very rapidly with increasing N. For clusters of identical atoms interacting with a pairwise potential, Wille and Vennik [2] have shown that the problem is "NP-hard" [3], i.e., there is no known algorithm that grows with time as power of N and with which one can determine the ground state energy and structure. The problems of calculating the energy are then *multiplied* with those of finding the most stable minimum.

Springer Series in Solid-State Sciences, Volume 114
Interatomic Potential and Structural Stability Eds.: K. Terakura and H. Akai

In the present lecture, we discuss results obtained using a method – the combination of density functional (DF) energy calculations with molecular dynamics (MD) – that addresses both aspects of the problem. DF calculations, with a local density approximation for the exchange-correlation energy, enable us to calculate energy surfaces efficiently, and MD allows us to determine stable structures in molecules and to follow the time-dependent changes in the structures of amorphous and liquid materials. We outline the main features of the method in Sec. II. In Sec. III we compare structures of clusters containing S and Se with structural properties of amorphous and liquid selenium, and results for phosphorus and arsenic clusters with the structures of amorphous and liquid phosphorus. The differences between the structures of elements of the same group (IIIa) are highlighted by the results of a study of aluminium and gallium clusters. The relationship to atomic properties is discussed in Sec. IV and our concluding remarks in Sec. V.

2. The molecular dynamics/density functional (MD/DF) method

The density functional (DF) formalism provides a numerically efficient scheme for calculating the *total energy* of the system of ions and electrons, without calculating the exact wave function. The basic theorems are [4]:

(1) Ground state (GS) properties (such as the total energy E) of the system in an external field V_{ext} can be determined from a knowledge of the electron density $n(\mathbf{r})$ *alone*.

(2) $E[n]$ satisfies the variational principle $E[n] \geq E_{GS}$, and the density for which the equality holds is the ground state density, n_{GS}.

The minimization of $E[n]$ leads to equations [5] of Hartree form, the solution of which leads to the energy and density of the lowest state, and all quantities derivable from them. An approximation to one part of the energy, the exchange-correlation energy E_{xc}, is presently unavoidable. In the present work we use the local spin density (LSD) approximation, which is free of adjustable parameters. For a detailed discussion of its applications, see the review article by Jones and Gunnarsson [6].

The problem of locating favourable structures amongst many possible is faced by crystal growers, who usually address it by increasing the temperature (often to the melting point) so that the kinetic energy of the atoms prevents the system becoming trapped in an energetically unfavourable structure. Slow cooling (annealing) then results in a stable structure. When carried out on a computer, this strategy is referred to as "simulated annealing" [7]. One way of performing this was suggested by Car and Parrinello [8], who showed that molecular dynamics (MD) could be combined with the density functional (DF) scheme for calculating total energies, without requiring any assumptions about ground state geometries. Finite temperature MD techniques allow an efficient sampling of the potential energy surface, and the LSD approximation in DF calculations avoids the parametrization of the interatomic forces common in

MD schemes. As with any MD approach, we can also study time-dependent properties of liquid and amorphous systems at non-zero temperatures.

3. Application to clusters of S, Se; amorphous Se

Sulphur and selenium provide some of the best characterized atomic clusters, since many allotropes comprise regular arrays of well-separated ring molecules, and X-ray structure analyses have been performed for S_n, $n = 6-8, 10-13, 18, 20$ [9,10] and Se_n, $n = 6, 8$ [9,11]. Several mixed crystals of the form Se_nS_m [12] are known, and their possible complexity is evident from the example of eight-membered rings, where 30 different crown-shaped isomers can occur. The preparation of all these molecules is reviewed by Steudel [10].

MD/DF calculations have been carried out for some of these systems, as well as for amorphous and liquid selenium. We use periodic boundary conditions, and the large unit cell in the cluster calculations [fcc lattice constant 15.9 Å] guarantees a very weak interaction between the clusters. In the case of the disordered systems we considered 64 Se atoms with a unit cell volume appropriate to the density of liquid Se at 573K [3.91 g cm^{-3}]. Further details of the calculations are found in Ref. [13].

3.1 Ring molecules containing S and Se.

The calculations for selenium [14] and sulphur clusters [13] showed that it is possible to determine low-lying energy minima even if one starts from a geometry that is far from optimal. In S_{3-6}, for example, we started from linear chains and in S_{7-13} from planar rings. After repeated heating and cooling cycles, the final structures at $T = 0$ agreed well with experiment in all cases where X-ray data were available. The D_{3d} symmetry of S_{12}, for example, is reproduced very closely, and the structural parameters (bond length, bond angle, and dihedral angle) are in good agreement with measured values. In the case of S_9, the crystallites that can be prepared are too small for X-ray diffraction analysis, so that details of the crystal structure is unknown. The calculated ground state structure (C_2 symmetry) is, however, consistent with measured Raman spectra, which indicate almost equal bond lengths and bond angles, as well as dihedral angles in the range $70-130°$.

The last result shows how trends in bond lengths, bond angles, and dihedral angles can be used to obtain structural information even in the absence of X-ray data. A study of all the known structures of sulphur clusters, both measured and calculated, showed some striking correlations [13]. The first is that the longest bonds in these molecules are always adjacent to short bonds. In fact, the bond length d is inversely related to the mean of the two neighbouring bonds. The second relationship is between the bond length d and the dihedral angle γ, where we find that d is shortest for dihedral angles near 90°.

Calculations on mixed ring molecules of S and Se [15] showed that bond lengths (d_{SS}, d_{SSe}, d_{SeSe}) and the corresponding bond strengths (E_{SS}, etc.)

transferred remarkably well between structures of the same family, e.g., Se_nS_{8-n}, and the S-S and Se-Se bonds had the same lengths as in S_8 and Se_8, respectively. This result enables us to predict the relative stabilities of the isomers of different families of mixed S-Se molecules [15].

3.2 Amorphous and liquid Se

Hohl and Jones [16] have performed MD/DF simulations at 350K and 720K to study the structural properties of amorphous and liquid Se, respectively.
As an example of the results, we show in Fig. 1 the bond angle distributions. The bond angles are distributed narrowly around the values found in crystalline modifications [106° in α-monoclinic Se, 103° in trigonal Se, 102° in chains constrained in mordenite crystals]. The dihedral angle distribution [Fig. 2] is much broader at both temperatures, reflecting the low energy cost of torsion in all modifications of S and Se. The distribution is essentially symmetrical around its least favourable value (0°) and peaks near 100°, where the lone pair repulsion between neighbouring atoms is a minimum. Both liquid and amorphous Se can be viewed as complex, branched chains with bond lengths and bond angles close to the crystalline values and with widely varying dihedral angles of varying phase. The importance of the dihedral angle in discussions of these largely twofold coordinated systems shows the limitations of model force fields that are restricted to two- and three-body forces.

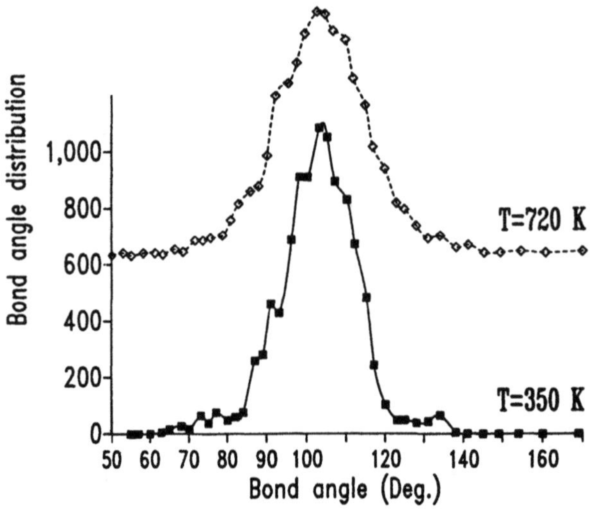

Fig. 1: Bond angle distribution in amorphous (squares) and liquid (diamonds) Se.

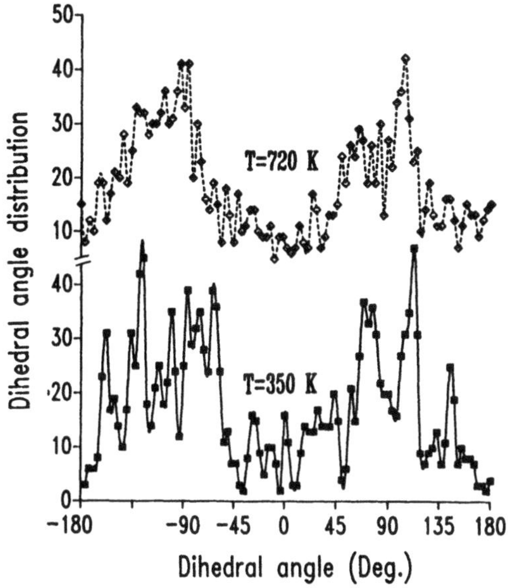

Fig. 2: Dihedral angle distribution in amorphous (squares) and liquid (diamonds) Se.

3.3 Phosphorus clusters up to P_{11}

MD/DF calculations on phosphorus clusters [17] have shown that both the number of stable minima and the structural variety increase with increasing N. The results are illustrated in Fig. 3 (a-d), where we show the most stable isomers of P_4, P_6, P_8, and P_{10}. The most stable form of P_4 is tetrahedral, with threefold rings and bond angles of 60°, so that all atoms are threefold coordinated. The most stable structure of P_6 has two threefold rings, which may be viewed as the "addition" of a dimer to P_4, as shown in Fig. 3(a-b). Threefold coordination is favoured in phosphorus clusters, but the number of threefold rings decreases with increasing cluster size. Such rings are virtually absent in crystalline allotropes of phosphorus. Larger rings – such as nonplanar fivefold rings – arise more frequently as the number of atoms in the cluster increases.

The average bond length in phosphorus clusters agrees well with those in both the crystalline and amorphous bulk systems. If we consider the structure of Hittorf's phosphorus (H-P) [Fig. 3(e)], we see long tubes of pentagonal cross-section containing P_8 and P_9 "cages", connected by pairs of phosphorus atoms [2,9]. Layers of such tubes are linked via bonds at the apex of the P_9 units. This unit was found to be one of the most stable structures in P_9 [8], and the P_{10} isomer with lowest energy is clearly related. It is quite remarkable that the "tubular" structure of H-P is already reflected in the stable structures of clusters with less than 12 atoms.

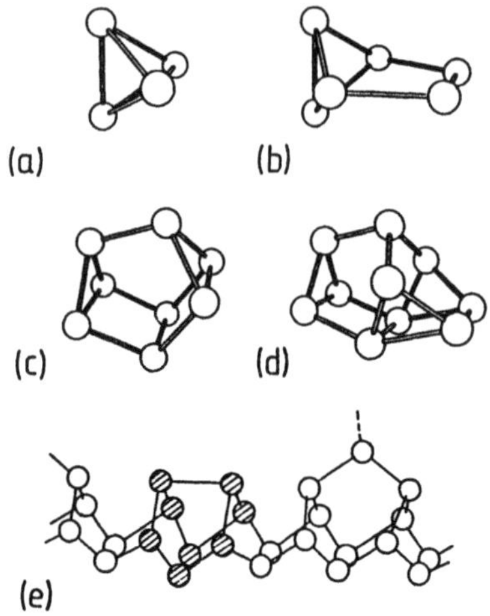

Fig. 3: Structures of the most stable structures of P_4, P_6, P_8, P_{10} compared with a chain subunit of Hittorf's phosphorus.

A parallel (unpublished) study of arsenic clusters has shown that the structures and relative stabilities of different isomers are very similar to those in phosphorus clusters. The main difference is an almost uniform increase in the bond lengths of $\sim 10\%$.

3.4 Amorphous and liquid P

The MD/DF simulations by Hohl and Jones [18] on amorphous phosphorus showed pronounced similarities to the trends discussed above. Threefold rings occur only rarely, and there are clear parallels with the structures of finite small clusters. As an example, we show in Fig. 4 the radial distribution functions found in experiment, two MD/DF simulations and in 70 clusters of up to 10 atoms. The similarities between the first three and the last are apparent, although the cluster results cut off due to their finite sizes. The structure of amorphous phosphorus found in the MD/DF simulations more closely resembles a network of cluster structures than any of the crystalline forms.

Simulations on liquid phosphorus are still in progress. An important question in this context is the polymerization at elevated temperatures of molten white phosphorus (comprising P_4 tetramers, with threefold rings) to the networks found in amorphous P. Present results indicate that the tetramer units open to give "roof" structures with two twofold coordinated atoms. These units can then interact to give structural units found in the amorphous state.

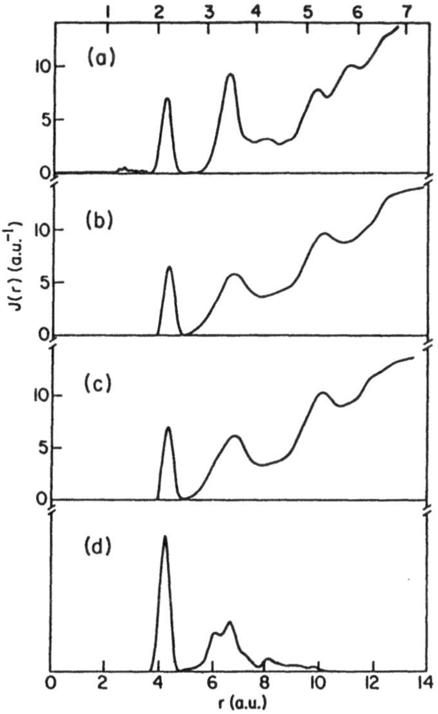

Fig. 4: Radial distribution functions $J(r)=4\pi r^2\rho(r)$ in amorphous P obtained from (a) experiment [19] (b,c) two simulation trajectories. (d) Histogram obtained from 70 structures of P_n clusters.

3.5 Clusters of Al and Ga

The final example we consider is the structures of small clusters of aluminium and gallium [20]. There is a rich variety of structures, with a transition from planar to non-planar structures at $N = 5$. The most stable structures of clusters with more than five atoms can be related to fragments of the f.c.c. structure, with structures being derived by capping stable isomers of smaller ones. The number of possible capped structures increases rapidly with increasing cluster size, so that this does not provide a unique prescription for determining the most stable isomers. For our present purposes, the most interesting result is that the bond lengths in the gallium clusters were *shorter* than those in the lighter element aluminium. This behaviour is different from those in the other main groups that we have considered, and will be discussed further below.

4. Relation of structural trends to atomic properties

We have noted above that Se-Se bonds are typically 12–15% longer than S-S bonds, and in group Va As-As bonds are 10% longer than P-P bonds. In both cases this trend can be related to the extent of the radial functions of the valence electrons, i.e., to the "size" of the atoms. These functions are shown for P and As in Fig. 5(a). The connection between bond length, bond strength, and the range of the atomic orbitals has been discussed in the DF context by Harris and Jones [21].

The results on group IIIa clusters, where Ga-Ga bonds are typically 3–7% shorter than Al-Al bonds, continue this trend. The plot of the radial functions of Al and Ga [Fig. 5(b)] shows that the tails of the valence p-functions are very similar, while the s-function of Ga is more compact. The relatively extended 3d-electrons in Ga do not screen completely the additional nuclear charge in the heavier element. While it is unusual that the bond lengths in a heavier element of a group (in this case Ga) are shorter than those of a lighter one, this effect is evident in the nearest neighbor separations of the bulk elements, 5.411 a.u. in f.c.c. Al and a weighted average of the seven nearest neighbours in Ga (5.107 a.u.).

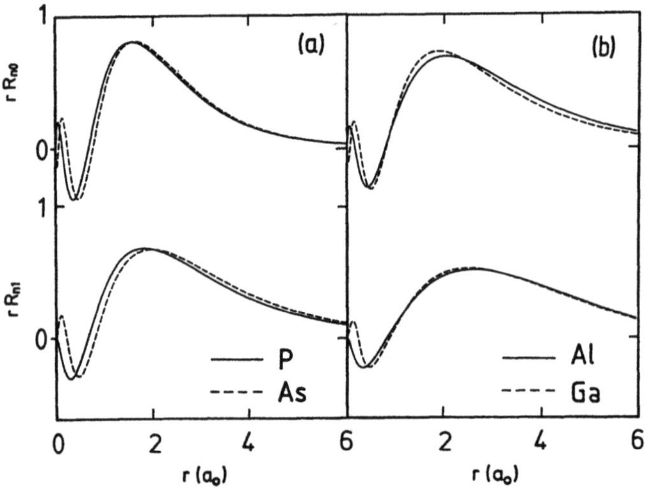

Fig. 5: Radial functions of s- and p-valence electrons in: (a) P, As, (b) Al, Ga.

5. Concluding remarks

The combination of molecular dynamics and density functional methods has been applied to small clusters, to mixed ring molecules of S and Se, and to liquid and amorphous materials. Experience has shown that the calculations lead to reliable results, reproducing known structures and predicting new ones. We have focused here on structural trends, both within different main groups and for different forms of the same element. We have seen that there are interesting parallels in all cases.

It is of particular importance that bonding trends are evident even in clusters with less than 10 atoms. The energy surfaces of such systems are much easier to study than those of liquids and amorphous materials, and the parallels indicate that forces derived from the energy surfaces of clusters should be useful in describing the forces in extended systems. In spite of the successes of the MD/DF method so far, however, it is important to recognize that *no* computational method can guarantee to find the absolute energy minima even in relatively small clusters.

I thank D. Hohl and G. Seifert for collaboration on parts of this work. The calculations were performed on Cray computers of the Forschungszentrum Jülich and the German Supercomputer Centre (Höchstleistungsrechenzentrum, HLRZ).

References

1. M.R. Hoare and J.A. McInnes, Adv. Phys. **32**, 791 (1983).
2. L.T. Wille and J. Vennik, J. Phys. A **18**, L419, L1113 (1985).
3. M.R. Garey and D.S. Johnson, *Computers and Intractability: A Guide to the Theory of NP-Completeness* (Freeman, San Francisco, 1979).
4. P. Hohenberg and W. Kohn, Phys. Rev. **136**, B864 (1964).
5. W. Kohn and L.J. Sham, Phys. Rev. **140**, A1133 (1965).
6. R.O. Jones and O. Gunnarsson, Rev. Mod. Phys. **61**, 689 (1989).
7. S. Kirkpatrick, C.D. Gelatt Jr., and M.P. Vecchi, Science **220**, 671 (1983).
8. R. Car and M. Parrinello, Phys. Rev. Lett. **55**, 2471 (1985).
9. J. Donohue: *The Structures of the Elements* (Wiley, New York, 1974), Chap. 9.
10. R. Steudel, in *Studies in Inorganic Chemistry, Vol. 5*, edited by A. Müller, B. Krebs (Elsevier, Amsterdam, 1984), p. 3.
11. R. Steudel and E.-M. Strauss, Adv. Inorg. Chem. Radiochem. **28**, 135 (1984).
12. R. Steudel, E.M. Strauss, in *The Chemistry of Inorganic Homo- and Heterocycles, Vol. 2*, (Academic, London, 1987), p. 769.
13. D. Hohl, R.O. Jones, R. Car, and M. Parrinello, J. Chem. Phys. **89**, 6823 (1988).
14. D. Hohl, R.O. Jones, R. Car, and M. Parrinello: Chem. Phys. Lett. **139**, 540 (1987).
15. R.O. Jones and D. Hohl, J. Am. Chem. Soc. **112**, 2590 (1990).
16. D. Hohl and R.O. Jones, Phys. Rev. B **43**, 3856 (1991).

17. R.O. Jones and D. Hohl, J. Chem. Phys. **92**, 6710 (1990); G. Seifert and R.O. Jones, Z. Phys. D **20**, 77 (1991); R.O. Jones and G. Seifert, J. Chem. Phys. **96**, 7564 (1992).
18. D. Hohl and R.O. Jones, Phys. Rev. B **45**, 8995 (1992).
19. S.R. Elliott, J.C. Dore, and E. Marsaglia, J. Phys. (Paris) **46**, Colloq. C8, 349 (1985).
20. R.O. Jones, Phys. Rev. Lett. **67**, 224 (1991).
21. J. Harris and R.O. Jones, Phys. Rev. A **19**, 1813 (1979).

Order-Disorder Phase Transition on the Si(001) Surface

K. Inoue[1], *Y. Morikawa*[2], *K. Terakura*[2], and *M. Nakayama*[1]
[1]College of General Education, Kyushu University, Ropponmatsul, Fukuoka 810, Japan
[2]Institute for Solid State Physics, University of Tokyo, Roppongi, Minato-ku, Tokyo 106, Japan

Abstract. On the Si(001) surface, atoms form asymmetric dimers, which are arranged in a c(4 × 2) pattern at low temperatures. As the temperature increases, the ordered phase transforms into a disordered phase around 200 K. This structural phase transition was studied theoretically by using a two-dimensional Ising spin model. The interaction parameters between Ising spins were estimated by the first-principles total energy calculations for some different arrangements of the asymmetric dimers. Monte Carlo simulations were then performed. Theoretical results account well for the experimental observations. Dramatic effects of dimer defects on the nature of the phase transition are also discussed.

1. Introduction

The bulk terminated ideal surface of Si(001) has a high surface energy because of the presence of two dangling bonds per surface atom. In order to reduce the surface energy, the surface Si atoms dimerize and form dimer rows. There is still one dangling bond per surface atom and the two dangling bonds of a dimer form a bonding π and an antibonding π^* states. These π and π^* states form π and π^* bands due to the interaction between the neighboring dimers. If the dimers are symmetric and nonmagnetic, the π and π^* bands overlap energetically and the surface will be metallic. Experimentally, however, the surface seems to be insulating [1]. Two possibilities were suggested theoretically to explain the experimental observation. One is the asymmetric dimer model first proposed by Chadi [2] and another is the antiferromagnetic order model [3,4]. The clear c(4 × 2) pattern below 200 K observed by LEED [5,6] may be favorable to the asymmetric dimer model, because the magnetic order may not be so clearly visible by spin-unpolarized LEED. On the other hand, the STM image at room temperatures [7,8] seems to suggest that the dimers are symmetric on the most part of the flat surface. This fact was taken as one of the evidences for the antiferromagnetic ordering. However, the recent STM measurement at low temperatures (< 200 K) [9] clearly ob-

Springer Series in Solid-State Sciences, Volume 114
Interatomic Potential and Structural Stability   Eds.: K. Terakura and H. Akai

served asymmetric dimers which orders in a c(4 × 2) pattern in a wide range of the surface. It is therefore believed that the room-temperature STM image may correspond to the time average of the flip-flop motion of the asymmetric dimers.

A detailed experimental study of the order-disorder phase transition was recently performed [6]: temperature dependence of the intensity profile for the quarter-order spot characteristic of the c(4×2) structure was carefully studied. The presence of a streak pattern associated with the quarter-order spot even at the temperatures twice as high as the order-disorder transition temperature T_c suggests a strong short range order along the dimer row even at such high temperatures.

A theoretical study of this phase transition was performed by Ihm et al.[10] and Saxena et al.[11]. They adopted an Ising spin model where the two degrees of freedom of an Ising spin correspond to the two possible orientations of an asymmetric Si-Si dimer. The interaction parameters in the model were estimated by using a tight-binding approximation for describing the electronic structure of the Si surface. In this analysis, the ground state configuration was p(2 × 2) rather than c(4 × 2).

The purpose of the present work is to refine the above theoretical works. The recent progress in the electronic structure calculation enables us to perform a large scale calculations with very high accuracy. The present pseudopotential calculation gives the correct ground state and provides us with a refined set of interaction parameters in the Ising spin model. A Monte Carlo simulation is performed for this model and several aspects of the order-disorder transition are studied. The calculated LEED intensity profile can explain the experimental observation [6] fairly well. However, a noticeable difference exists between theory and experiment with respect to sharpness of the phase transition . Although the theoretical order parameter for the c(4 × 2) pattern indicates a sharp second order phase transition at the critical temperature T_c, the experimental LEED intensity for the quarter order spot shows only a gradual change around T_c [6]. We point out that the presence of only a small amount of dimer defects ($\sim$ 1%) on the real surface may be responsible to this discrepancy. Defects in the present problem are equivalent to random magnetic fields in an Ising spin system: a rigorous theory proves the non-existence of the ordered phase for any weak random magnetic field [12,13]. The very dramatic effect of dimer defects on the phase transition in our Monte Carlo simulation is consistent with this rigorous theory.

The calculational procedure is briefly described in Section 2, results and discussion are given in Section 3 and we conclude in Section 4.

2. Calculational procedure

By regarding the two possible orientations of an asymmetric Si-Si dimer as the two degrees of freedom of an Ising spin, we assume the following Ising spin Hamiltonian in order to discuss the order-disorder phase transition of the c(4 × 2) pattern on the Si(001) surface.

$$H = -V\sum_{i,j} S_{i,j}S_{i,j+1} - H\sum_{i,j} S_{i,j}S_{i+1,j} - D\sum_{i,j} S_{i,j}(S_{i+1,j+1} + S_{i+1,j-1}). \quad (1)$$

The exchange interactions V, H and D are defined as indicated in Fig. 1 and their actual values are derived from the total energy calculations for the four different configurations of the asymmetric dimers: (2 × 1), (4 × 1), p(2 × 2) and c(4 × 2) as shown in Fig. 1a of ref.10. The procedure is essentially the same as that of Ihm et al.[10] except the method of the total energy calculation. While Ihm et al. adopted a tight-binding approximation, we adopt the norm-conserving pseudopotential [14] with the plane-wave expansion for the eigenfunctions. Our calculation is based on the local density approximation in the density-functional theory and the structural optimization is performed by the first-principles molecular dynamics method [15].

The computational conditions in the present work are summarized as follows. In the repeated slab calculation, one slab unit consists of 12 Si layers and a vacuum region of 10 Si-layer thickness. The cutoff energy for the plane-wave expansion is 12.25 Ry and the numbers of sampling

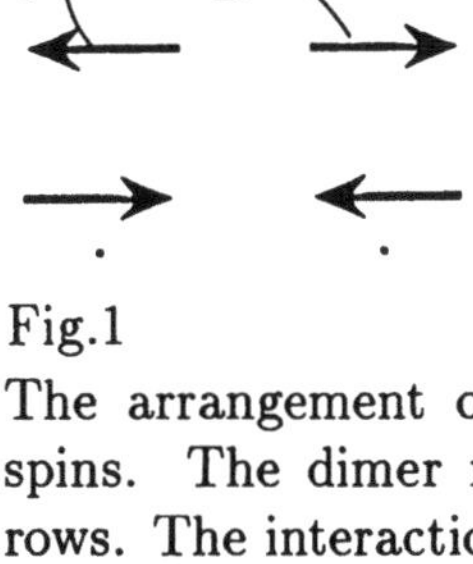

Fig.1
The arrangement of asymmetric dimers represented by the Ising spins. The dimer rows are perpendicular to the directions of arrows. The interaction parameters V, H and D in Eq.(1) are defined as shown in the figure.

k-points in the irreducible part of the Brillouin zone are 8 for (2×1) and 4 for other three structures. After optimizing each structure, the total energies of the four structures are calculated with a common (4×2) unit cell. At this final stage of the total energy comparison, two k-points are used for all the four structures. However, as the energy difference between $c(4 \times 2)$ and $p(2 \times 2)$ is very small, calculations with eight k-points are also performed for checking the reliability of the result. More details of the calculational procedure were described in our previous publications [16,17] and will also be described in a separate paper [18].

Once the values for V, H and D are obtained, a standard Monte Carlo simulation with the Metropolis algorithm is performed with the Hamiltonian of Eq.1. As the interaction V is strong, the correlation length along the dimer row is expected to be long. Therefore, our system contains 20 dimer rows each of which contains 500 dimers along it. The temperature is lowered starting from 600 K far above the critical temperature T_c. For every given temperature, the first 10^4 Monte Carlo steps are simply regarded as the annealing process and the following 10^4 (in most cases) or 7×10^4 (near the critical temperature) Monte Carlo steps are used for measuring thermodynamic quantities. In this measuring Monte Carlo process, one data point is obtained for every 10^3 or 7×10^3 Monte Carlo steps and 10 such data points are finally used to estimate the standard variations, which are shown as error bars in the figures presented later.

On the real Si(001) surfaces, appreciable amount of dimer defects exist as demonstrated clearly by the STM image [7-9]. We study the effect of dimer defects in two phenomenological ways. In the first one, a dimer defect is simply regarded as missing of a spin at a particular site. Therefore, such a defect causes decoupling of interactions between spins around the defect. We call this a free-spin defect. In the second model, a dimer defect prohibits the flip-flop motion of its nearest neighbor dimers along the dimer row. In the present study, the two spins next to the defect are forced to be parallel each other and two possible defects corresponding to two possible orientations of the neighboring spins are distributed randomly. This kind of defects are called fixed-spin defects. In both cases, the defect concentration is assumed to be 1%.

3. Results and Discussion

The total energies per Si dimer for the four different asymmetric-dimer arrangements are shown in Table 1 with the total energy for the $c(4 \times 2)$ structure as the reference. The (4×1) and (2×1) structures are definitely higher in energy than the $p(2 \times 2)$ and $c(4 \times 2)$ structures, while the $c(4 \times 2)$

Tabel 1: Total energies for four different reconstruction geometries with reference to the c(4 × 2) structure and the coupling constants decuced from them. The values in the parentheses are the coupling constants calculated by Ihm et al[10].

	E (meV)	
(2x1)	90.6	
(4x1)	117.6	
p(2x2)	1.2	
c(4x2)	0.0	
V	-51.9	(-26)
H	6.6	(10)
D	3.6	(4)

structure is only marginally lower in energy than the p(2 × 2) structure. One may think that the energy difference of 1.2 meV/dimer will be at or outside the limit of numerical accuracy. However, a check calculation with eight k points gives 1.4 meV/dimer as this energy difference. Some additional test calculations with smaller cutoff energies do not change the sign at all and give always a similar magnitude. Although Ihm et al.'s tight-binding calculation [10] led to a wrong energy ordering between the p(2 × 2) and the c(4 × 2) structures, the present calculation predicts the ground state correctly.

From the total energies listed in Table 1, the three interaction parameters V, H, and D are obtained and listed also in Table 1. The figures in the parenthesis are Ihm et al.'s result [10]. We would like to point out two important aspects. First, our V, the interaction along the dimer row is twice as large as Ihm et al.'s value. Second, the sign of $H - 2D$ determines which of the c(4 × 2) and p(2 × 2) structures has a lower energy. The present value of $H - 2D$ is negative and stabilizes the c(4 × 2) structure.

The order parameter for the c(4 × 2) structure is shown in Fig. 2. At $T = 320$ K, the order parameter drops sharply and the susceptibility (not shown here) has a sharp peak. Therefore, the critical temperature T_c in the present theory is 320 K, which is slightly higher than the experimental value of about 200 K [6]. Nevertheless, considering that the present calculation does not contain any adjustable parameter, we think that the theoretical T_c is in good agreement with the experimental one.

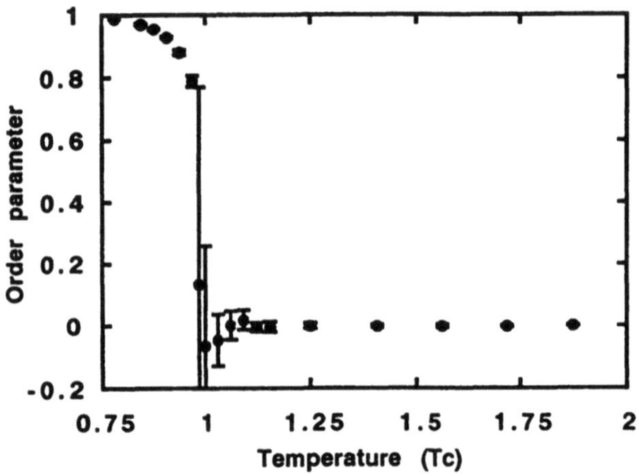

Fig.2
The temperature dependence of the order parameter for the c(4 × 2) structure for a perfect Si(001) surface. The temperature is measured in units of $T_c = 320$ K.

The LEED intensity profile calculated by a kinematical theory is shown in Fig. 3 for three different temperatures. Below the critical temperature (T = 300 K, T/T_c = 0.938), a sharp quarter-order peak is superposed with a streak which is sharp (broad) in parallel (perpendicular) to the dimer row. Above the critical temperature (T = 400 K, $T/T_c = 1.25$), the streak becomes broader in perpendicular to the dimer row but still peaks up at $(\frac{3}{4}, \frac{1}{2})$ in the k space. Even at a temperature twice as high as T_c ($T = 600$ K, $T/T_c = 1.875$), a streak still exists. All of these aspects can be seen in the experimental data [6] if the temperature is scaled with T_c.

A qualitatively different aspect of the present theory from the experiment can be seen in the sharpness of the phase transition. While the order parameter in Fig. 2 indicates a sharp phase transition at $T_c(= 320$ K), the experimental LEED intensity has a broad transition range around 200 K. We want to point out that the presence of dimer defects on the real surfaces may be responsible for the diffuseness of the phase transition. Figure 4a shows the order parameter for the c(4 × 2) structure on the surface with 1% free-spin defects. In this case, the dimer defects are assumed to cut the interaction with neighboring dimers. Therefore the flip-flop motion of buckled dimers are thermally activated more easily. Comparing Fig. 4a with Fig. 2, we observe that this kind of defects reduce the critical temperature slightly without changing the basic nature of the phase transition.

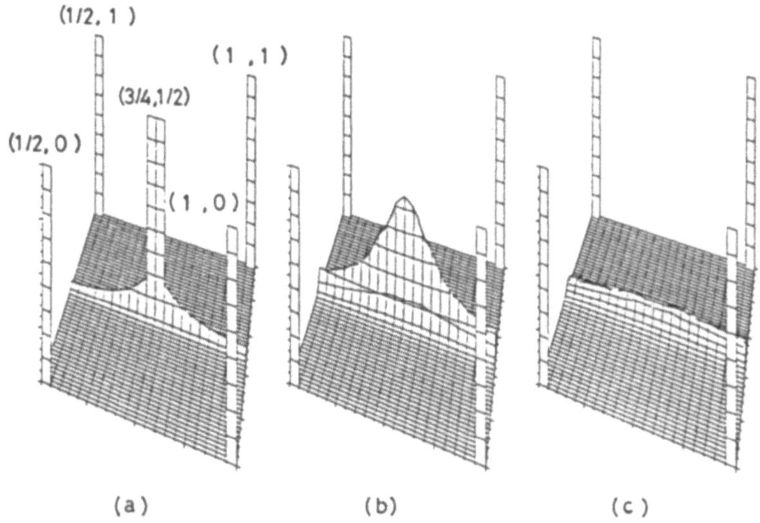

Fig.3
The kinematical LEED intensity curves for a perfect Si(001) surface.
(a) 0.938 T_c, (b) 1.25 T_c and (c) 1.875 T_c with $T_c = 320$ K.

Completely different behavior of the order parameter is realized by 1% fixed-spin defects as demonstrated by Fig. 4b. It is remarkable that only 1% defects reduces the order parameter by about 60% and that the transition becomes very diffuse. This result had been puzzling to us before we noticed that the present system is equivalent to an Ising spin system with a random magnetic field. According to a rigorous theory, a long-range magnetic order does not exist by non-vanishing strength of a random magnetic field for the system dimension d less than and equal to two [12,13]. In our fixed-spin defect model we are assuming that an infinitely strong local magnetic field is applied at the spins next to the defect. In order to see the details of the asymmetric dimer arrangement for 1% fixed-spin defects, we show in Fig. 5 a snap shot of a part of our system in the Monte Carlo simulation at 250 K ($T/T_c = 0.78$). A couple of comments may be helpful to understand Fig. 5.

i) In the c(4 × 2) structure, the spin variable can be expressed either by $S_{i,j} = (-1)^{i+j}$ or $S_{i,j} = -(-1)^{i+j}$. These two cases correspond to two different domains of the c(4 × 2) structure, which are distinguished by the yellow and blue regions. We call them +(−) domain for short.

ii) Dimer rows are arranged horizontally. Only the raised atom of an asymmetric dimer is denoted by a sphere. Therefore, within a domain of the c(4 × 2) structure, spheres form a honeycomb lattice.

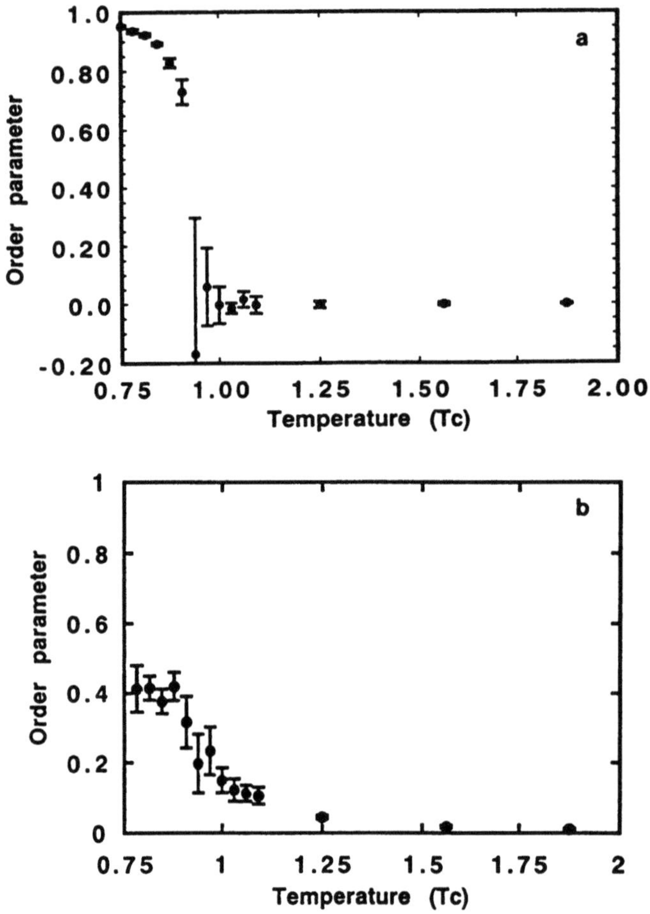

Fig.4

The temperature dependence of the order parameter for the c(4 × 2) structure for a Si(001) surface. (a) for 1% free-spin defects and (b) for 1% fixed-spin defects. The temperature is measured in units of $T_c = 320$ K.

iii) Along dimer rows, one may notice some places where a sphere is missing. These places correspond to dimer defects.

iv) Along a dimer row, the dimers next to a dimer defect are not allowed to do thermal flip-flop motion.

In the present model for fixed-spin defects, a single dimer defect can stabilize the c(4 × 2) structure. However, if we introduce more than two dimer defects, effects of interference between them become important. Each of dimer defects tends to stabilize the c(4 × 2) structure locally.

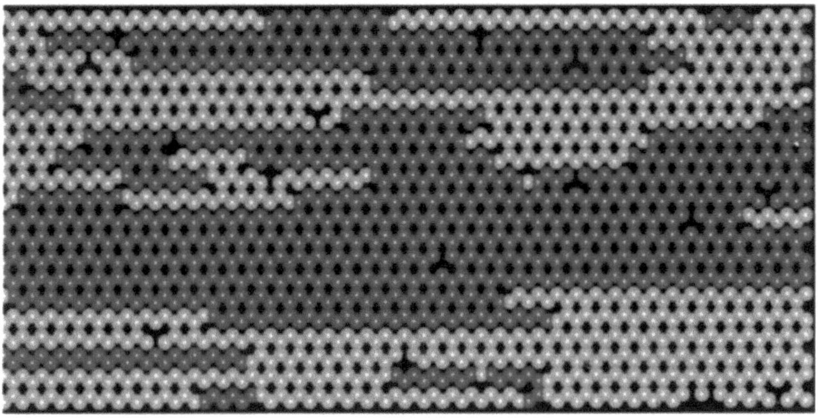

Fig.5
A snap shot of a part of the system with 1% fixed-spin defects in the Monte Carlo simulation at 0.78 T_c with $T_c = 320$K. The dimer rows run horizontally and only the raised Si atoms of asymmetric dimers are denoted by spheres. The yellow and blue regions correspond to two domains characterized by $S_{i,j} = (-1)^{i+j}$ and $S_{i,j} = -(-1)^{i+j}$, respectively.

These local c(4 × 2) structure may be either + or – domain. If mutual relation among defects in a part of the system is consistent with the formation of a specific c(4 × 2) domain, this part of the system may have a stable c(4 × 2) domain. Otherwise, fixed-spin defects tend to divide the system into several domains. Therefore the order parameter in our system is drastically reduced. As mentioned already, the order parameter must vanish in the thermodynamic limit of the system size. It means that the finite order parameter in Fig. 4b comes from fluctuation in our limited system size. In the real LEED experiment, the coherence of the incident beam is kept only within a region with a linear dimension of some hundreds Å, which is comparable with our system size. Therefore, the quarter-order LEED spot characteristic of the c(4 × 2) structure has the same origin as the order parameter of Fig. 4b. The broad transition region observed experimentally can thus be explained by the presence of fixed-spin defects on the real surface. The Monte Carlo step sequence of the asymmetric dimer arrangement revealed that near the domain boundaries introduced by fixed-spin defects, the flip-flop motion is fairly frequent even at $T = 250$ K ($T/T_c = 0.78$). On the perfect surface without defects, however, the whole region of the system forms a single stable c(4 × 2) domain and the flip-flop motion of dimers occurs only occasionally. It is interesting to note that on the surfaces both with and without defects a single-dimer flipping is accompanied with flipping of dimer along a dimer row and sometimes the flipping-motion propagation extends to more than 20 dimers.

4. Conclusion

The order-disorder phase transition of the Si(001)-c(4 × 2) surface was theoretically analyzed. The first-principles molecular dynamics approach to the total energy calculations for some different arrangements of asymmetric dimers successfully reproduced the c(4 × 2) structure as the lowest configuration. From the calculated total energies, the interaction parameters in the model Ising spin Hamiltonian of Ihm et al.[10] were derived. The Monte Carlo simulation was then performed and the critical temperature was estimated to be $T_c = 320$ K. Considering that the present calculation is totally parameter free, the above critical temperature agrees reasonably well with the experimental one of about 200 K. The temperature dependence of the LEED streak pattern around the quarter order spot characteristic of the c(4 × 2) structure can also be reproduced fairly well. In particular, the streak pattern persists up to the temperatures twice as high as the critical temperature both theoretically and experimentally.

However, the theoretical result and the experimental one are not consistent with each other with regard to the sharpness of the phase transition. Experimentally the transition is very broad, while it is fairly sharp theoretically. It was demonstrated that a small concentration of dimer defects on the real surface can be the cause of such a broad transition. The dimer defects in the present treatment are equivalent to a random magnetic field in the two-dimensional Ising spin system. It was rigorously proven that the long range order does not exist in the latter system [12,13]. Therefore, the experimental LEED intensity for the quarter-order spot may come from fluctuation in the finite coherence range of the incident electron beam. As the linear-dimension of the coherence range is some hundreds Å, the calculated order parameter for a finite system with a size given by 500 × 20 may have a good correspondence with the observed LEED intensity.

Acknowledgement

The authors are grateful to Prof. Y. Murata for valuable discussions in several aspects of the present work. They are indebted to Prof. M. Imada for pointing them the equivalence of the present defect problem with the random magnetic field problem in the two-dimensional Ising spin system. Many useful comments from Dr. Y. Sakamoto on the Monte Carlo simulation are gratefully appreciated. We also thank Dr. T. Yamasaki for his invaluable contribution to making video for the present Monte Carlo simulation.

The numerical calculations were performed at the computer centers of University of Tokyo, Hokkaido University, Institute for Solid State Physics, and Institute for Molecular Science. This work was partially supported by a Grant-in-Aid for Scientific Research on Priority Areas from Ministry of Education, Science, and Culture.

References

1 F. J. Himpsel and D. E. Eastman, J. Vac. Sci. Technol. **16**, 1297 (1979).
2 D. J. Chadi, Phys. Rev. Lett. **43**, 43 (1979).
3 È. Artacho and F. Ynduráin, Phys. Rev. Lett. **62**, 2491 (1989).
4 E. Artacho and F. Ynduráin, Phys. Rev. B **42**, 11310 (1990).
5 T. Tabata, T. Aruga and Y. Murata, Surf. Sci. **179**, L63 (1987).
6 M. Kubota and Y. Murata, to be published in Phys. Rev. B.
7 R. M. Tromp, R. J. Hamers and J. E. Demuth, Phys. Rev. Lett. **55**, 1303 (1985).
8 R. J. Hamers, R. M. Tromp and J. E. Demuth, Phsy. Rev. B **34**, 5343 (1986).
9 R. A. Wolkow, Phys. Rev. Lett. **68**, 2636 (1992).
10 J. Ihm, D. H. Lee, J. D. Joannopoulos and J. J. Xiong, Phys. Rev. Lett. **51**, 1872 (1983).
11 A. Saxena, E. T. Gawlinski and J. D. Gunton, Surf. Sci. **160**, 618 (1985).
12 Y. Imry and S. Ma, Phys. Rev. Lett. **35**, 1399 (1975).
13 As a review article, D. S. Fisher, G. M. Grinstein and A. Khurana, Phys. Today Vol.41, No.12, 56 (1988).
14 G. B. Bachelet, D. R. Hamann and M. Schlüter, Phys. Rev. B **26**, 4199 (1982).
15 R. Car and M. Parrinello, Phys. Rev. Lett. **55**, 2471 (1985).
16 Y. Morikawa, K. Kobayashi, K. Terakura, and S. Blügel, Phys. Rev. B **44**, 3459 (1991).
17 K. Kobayashi, Y. Morikawa, K. Terakura, and S. Blügel, Phys. Rev. B **45** 3469 (1992).
18 K. Inoue, Y. Morikawa, K. Terakura and M. Nakayama, in preparation.

First-Principles Structural Optimization of the Al/Si Surface

T. Fujiwara[1], *Y. Yamamoto*[2], *M. Arai*[1], *and H. Tsuge*[3]

[1]Department of Applied Physics, University of Tokyo, Bunkyo-ku, Tokyo 113, Japan
[2]Hitachi Central Research Laboratory, Kokubunji, Tokyo 185, Japan
[3]Nippon Steel Corporation, Research and Developing Laboratory, Nakahara, Kawasaki 211, Japan

Adsorption of Al atoms on Si(111) and Si(100) surface is studied using the first-principles molecular dynamics method. For higher coverage of Al atoms, a new uniform metallic phase of an adsorbed film appears, though at lower coverage the dimer formation stabilizes the surface structure. A new algorithm without any diagonalization is also proposed, which is appropriate for degenerate Fermi levels in metallic systems.

1. Introduction

Determination of atomic positions and electronic structures in condensed matters is the first step towards understanding their physical properties. It has required much computational efforts to determine the atomic structures in solids from a theoretical view point and still much effort is being made to investigate very reliable and efficient methods. In conventional methods, several plausible structures are assumed, and the total energies must be calculated. The structure of the lowest energy may be accepted as a candidate of the ground state structure, but not all possible structures could be studied.

Recently Car and Parrinello proposed new simulation technique[1], which is now called the first-principle molecular dynamics method or simply the Car-Parrinello method (we refer to this as the C-P method hereafter). In standard electronic structure calculations, electronic wavefunctions are obtained by solving the Kohn-Sham equation selfconsistently, equivalent to minimization of the LDA (Local Density Approximation) total energy functional, and diagonalizing the large LDA Hamiltonian matrix. In the C-P method, on the contrary, the electronic wavefunctions are regarded, as well as atomic coordinates, as *classical coordinates* whose potential energy is the LDA energy functional. An equation of motion of electronic wavefunctions is solved in an iterative way, introducing a ficti-

Springer Series in Solid-State Sciences, Volume 114
Interatomic Potential and Structural Stability Eds.: K. Terakura and H. Akai

tious electron mass, and, therefore, a diagonalization of large matrices is not required. Since the computational time of matrix diagonalization is proportional to N^3, if the size of a matrix is $N \times N$, this part is time-consuming in conventional methods. In the C-P method, for obtaining the eigenenergies and eigenfunctions, the iterative method for the equation of motion is used and is not the most time-consuming part. The most time-consuming part is a calculation of the LDA potential from the plane waves, proportional to $N \log N$. The most essential point in the C-P method is to perform both the simultaneous minimization of energies of the atomic and electronic parts and the annealing processes concurrently. Recently it has been more common to use, instead of the equation of motion method, the steepest decent method or the conjugate gradient method for optimizing the ground state structure [2]. The atomic forces are obtained by the first-principle electronic structure calculation. The force acting on atoms is a sum of the Hellmann-Feynman force and the Pulay force due to the incompleteness of basis functions in a practical calculation [3]. It is also essential in the C-P method to use the plane wave basis set as the electronic wavefunctions because the Pulay force vanishes in case that the basis set of electronic wavefunctions is independent on the atomic coordinates.

The invention of the norm-conserving pseudopotential method is very much crucial to the C-P method, because the plane wave basis set is very much appropriate for the force calculation. However, the applicability of the norm-conserving pseudopotential is restricted because of quite large number of plane waves necessary for the first-row elements and transition metal elements, and a more widely applicable first-principle pseudopotential is now much appreciated. The ultra-soft pseudopotential[4] and also the C-P method associated with the APW method[5] are very promising along this line. Furthermore, several efforts would be necessary to applied the C-P method to metallic systems, because the degeneracy at the Fermi energy may induce an abrupt change of electronic charge density during the simulation and it would cause numerical instability.

We will briefly review recent works of our group in the field of the first-principle structural optimization using the C-P method. It covers (1) the Si(111)-$(\sqrt{3} \times \sqrt{3})R30°$-Al phase of 1/3 coverage [4], (2) the Si(100)-Al interface structures with coverage of $\Theta = 1$ and 4/3 [5] and (3) a stable algorithm using fractional occupations and finite temperatures applying to metallic systems [6].

2. Si(111)-$(\sqrt{3} \times \sqrt{3})R30°$-Al

Interface structures Al/Si have attracted much attention as a prototypic metal-semiconductor interface and have been studied experimentally using several techniques such as the low electron diffraction (LEED) or the scanning tunnelling microscopy (STM). Very recently, a new experimental technique, the Ion Cluster Beam method has been developed which enables a direct growth of Al layers on a Si substrate without any transition layers. This technique allows a realization of almost perfect growth, without any transition layer, of Al(111) on Si(111) or Al(110) on Si(100). [9] These are motivation of our research of the interface structure of Al/Si.

The first is an isolated Al atoms on Si(111) surface. There are two possible absorption sites on the Si(111) surface; one is called T_4 site and the other H_3 site. Both are the three-fold symmetric sites and there is a Si atom two layer just below down the T_4 site but four layer below down the H_3 site. Experimental results favor a T_4 site [10]. We calculated the relaxed surface structure for both possibilities, T_4 and H_3 sites.

In the present section and the next, we use one **k**−point (Γ−point) within the reduced Brillouin zone of the extended unit cell and the cut-off energy of the plane-wave basis is set to be 6.0 Ry, which results in about 3000 plane-waves. The norm-conserving pseudopotentials of Hamann, Schlüter and Chiang [11] with the separable form of Kleinman and Bylander [12] is used and the exchange-correlation energy is calculated by using the Ceperly-Alder functional form [13]. The separable form of the norm-conserving pseudopotential with the LDF potential reproduces cohesive properties of Si and Al quite accurately. Prior to structural relaxation, the LDF energy functional is minimized first only with respect to the electronic orbitals to get an electronic ground state for initial atomic positions. Then the total energy functional is minimized by the steepest-descent method with respect to both atomic and electronic structures. Atoms will be moved by amounts proportional to the force until the maximum absolute value of forces becomes less than 1.5×10^{-3}a.u./a.u..

The periodicity perpendicular to the surface is assumed with four double layers of Si atoms and $\sim$ 2.6 double layer thickness of vacuum, totally 18.80Å. The supercell size used in the calculation is $11.51 \times 6.65 \times 18.80\text{Å}^3$. Then 48 Si atoms and 2 Al atoms in a unit cell are involved, among them four Si layers are fixed at the positions of the bulk structure. Al atoms are put from the above.

We found that the T_4 site is stabilized than the H_3 site by 0.56eV per Al adatom. The final atomic positions are shown in Fig.1. Detailed information was published elsewhere.[6] In the present calculation, we did not adopt any symmetry restriction except a shape of the unit cell and, therefore, the resultant atomic positions slightly violate the desired symmetry. When Al atoms are put on the H_3 sites, the displacement of Si atomic positions of the ideal surface is very small. In case of Al atoms on the T_4 sites, the Si atoms of the second layer just below the Al

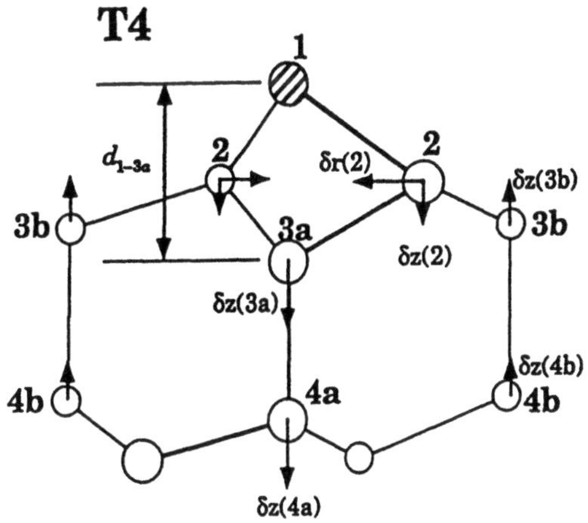

	(Å)
δr(2)	0.16,0.11
δz(2)	-0.09,-0.02
δz(3a)	-0.37
δz(4a)	-0.25
δz(3b)	0.11
δz(4b)	0.06
$d_{1\text{-}3a}$	2.50

Fig.1: Side view of the stable structure of Si(111)-($\sqrt{3} \times \sqrt{3}$)$R30°$-Al.

atoms shift downward and other Si atoms on the same layer move upward. These displacements allow the first layer Si atoms closer to Al adatoms and enhance the bonding charge density between Al atoms and the Si substrate.

3. Al/Si(100) interface structures

The new experimental technique, Ionized Cluster Beam (ICB) method [9] realizes an epitaxial growth of Al(111) films on Si(111) or Al(110) on Si(100) without transition layers, though lattice constants of Si and Al are different more than 25%. Because a low concentration of Al adsorption atoms produces Al dimers on Si surfaces, we presume that the high concentration adsorption is far different from the lower concentration adsorption. One possibility may be that the growth of Al(110) layer on Si(100) surface with Al[001] direction parallel to Si[011] direction allows an epitaxial growth, because a commensurate matching of the lattice parameters may be established within 5% in such way as $(3/2)a_{\mathrm{Si}[011]} \simeq (4/2)a_{\mathrm{Al}[011]}$, and $(1/2)a_{\mathrm{Si}[011]} \simeq a_{\mathrm{Al}[100]}$. We simulated this possibility by the C-P method. It may be very crucial that the kinetic energy of clusters are essential to elevate local temperatures and induce diffusion process of Al atoms near the surface, though our simulation is at the zero temperature with no kinetic energy and searching a possible local energy minimum.

The unit cell used in the calculation consists of six Si(100) layers of 7.66×11.44Å^2 in size, each containing six Si atoms, and the vacuum with twice thickness of Si layers, totally 20.317Å thickness, and the periodic boundary condition is imposed. The surface size of Si(100) layer contains

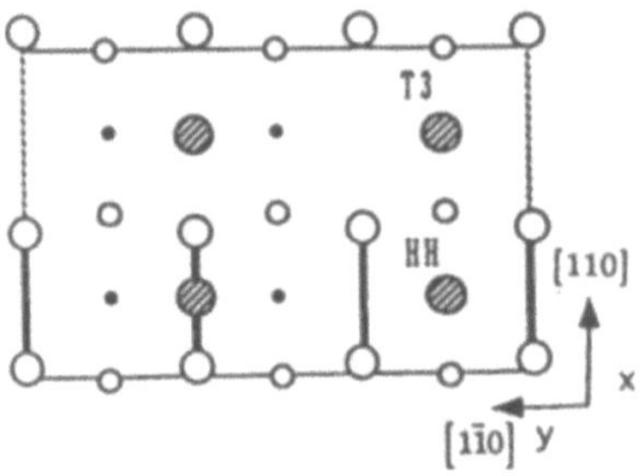

Fig.2: Possible adsorption sites T3 and HH on Si(100).

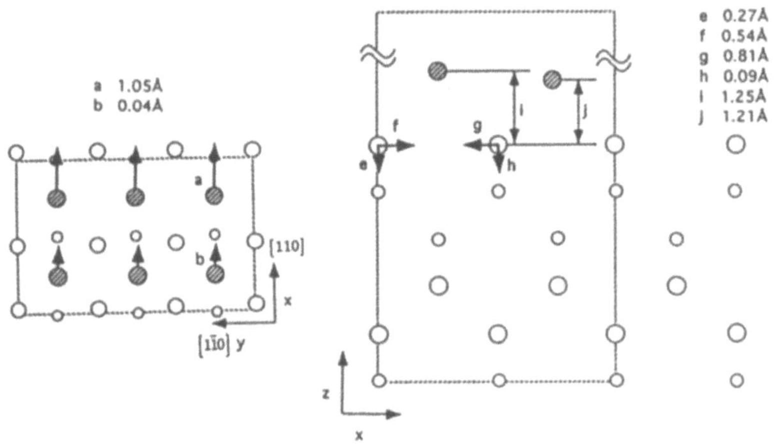

Fig.3: Optimal adsorption site at $\Theta = 1$ and displacements from the idealized positions.

actually six 2-dimensional unit cells of the ideal (100) surface and the size perpendicular to the surface is of 15 Si layers. This is the minimum size in order to see the proposed commensurate structure of Al-Si(100) interface. We put 6 and 8 Al atoms on the Si surface and optimized their positions along with the positions of Si atoms within topmost three layers, and the lower three Si layers are fixed at the ideal bulk configuration.

First, we studied the behavior of adsorbed Al atoms when the coverage $\Theta = 1$, i.e. six Al atoms on the T3 and HH sites (three each on T3's and HH's), and optimized the structure. The T3 and HH are the equivalent sites on an ideal (unreconstructed) surface. On a real clean surface, the topmost layer are the one of reconstructed asymmetric dimers and HH sites are on an array of dimes as shown in Fig.2. A convergence required about 1200 iterations for $\Theta = 1$ cases. Al atoms made sets in two's, no bond charge between Al and Si, and formed dimers. The one-electron density is low in the region between dimers, and each dimer is isolated. Figure 3 shows the equilibrium atomic positions and Al dimer bonds (between Al

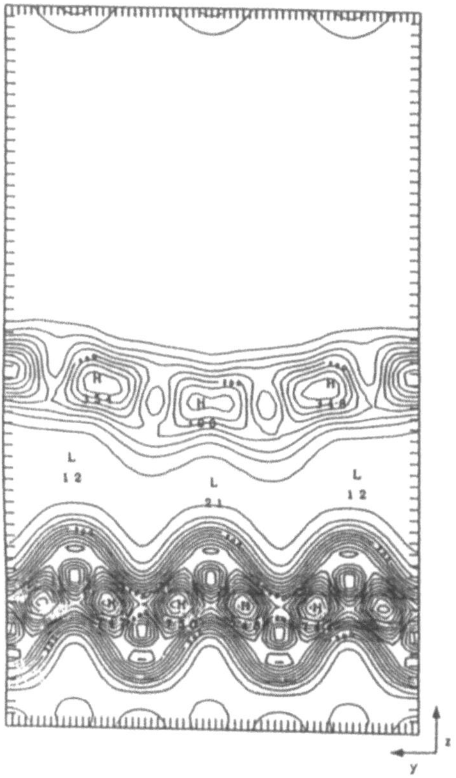

Fig.4: One electron density of the $\Theta = 4/3$ within a plane normal to Si surface and containing four Al atoms.

atoms on T3 and HH sites) are parallel to that of Si dimers. The shift of Al on T3 site is much larger than that on HH site. The bond length of Al dimer is 2.83Å which is almost the same as the nearest neighbor Al-Al distance in fcc Al, and that of Si dimer is 2.50Å.

For a coverage $\Theta = 4/3$ (8 Al atoms in a unit cell), we carried out calculation with two different initial configurations but the resulting atomic configurations are more or less similar, the height of eight Al atoms became uneven and the flat topmost Si surface is not retained. Iteration times necessary for convergence grew rapidly and is more than 2000. When projected Al atomic positions on the Si surface, the optimized arrangements of Al atoms are very like that of Al(110) in both two different initial configurations. The distance between adjacent Al atoms is almost equal everywhere, and no dimer formation was seen. The distance between the Al layer and the topmost Si layer is 2.5~3.3Å, which is two or three times larger than the lower coverage cases. Of course, this is quite natural because some Al atoms locate just above Si atoms. The charge distribution at this coverage also resembles that of solid Al, the valence charge spreads almost uniformly in the Al layer, and the bonds between Al and substrate Si are rather weak. This bond state is essentially different from those at lower coverage. We show in Fig.4 the one-electron density within a plane containing four Al atoms and normal to the Si surface. We also tried to simulate 16 Al atoms put on Si surface and, though we could get only a less convergence, two Al layers stacked on the Si surface.

4. Numerically stable algorithm using fractional occupancies for metallic systems

In the case of insulators and semiconductors with a band gap, the algorithm of the C-P method has been already established. However, some difficulties would arise in case of metallic systems because of degenerate states at the Fermi level.[14] First, the correct LDA ground state in metals has partially occupied degenerate levels at the Fermi energy, but the original C-P method cannot automatically reproduce it and produces a low symmetric state if integer occupation numbers are used. Second, during the course of the calculation, the level-ordering of occupied and unoccupied states may change, which causes sudden change of the occupation numbers at the Fermi level and the charge density in the real space. However, the original C-P method cannot follow this abrupt change very faithfully and brings the system away from its Born-Oppenheimer surface.

These difficulties may be overcome if fractional occupation numbers are introduced. In this scheme, the total energy functional is no longer invariant under a unitary transformation between the orbitals, and the correct representation is the one in which the effective Hamiltonian matrix is diagonal. So, the diagonalization process during the course of energy minimization has been recommended. This spoils one of the advantages of the C-P method and may be troublesome when we treat larger systems. Furthermore, this would disturb the minimization process with instability of the convergence. We now propose the most reasonable extension that is a treatment of the fractional occupation numbers $\{f_i\}$ as independent variables in addition to the orbitals $\{\psi_i\}$ and a usage of the grand potential as a variational functional instead of the total energy functional. This procedure allows an natural diagonalization without an explicit diagonalization procedure and provides a stable scheme.

Mermin extended the the Kohn-Sham theory to finite temperatures [15], where the following grand potential functional is minimized with respect to a one particle density $n(\mathbf{r}) = \sum_i f_i |\psi_i(\mathbf{r})|^2$:

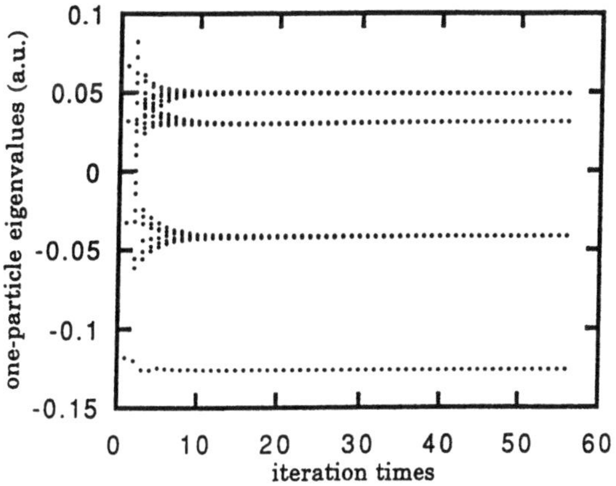

Fig.5: Convergence of one-electron energy of BCC Na.

$$
\begin{aligned}
&\Omega[\{\psi_i\},\{f_i\},\{\mathbf{R}_n\}] \\
&=\sum_i f_i \int \psi_i^*(\mathbf{r})(-\frac{1}{2}\nabla^2)\psi_i(\mathbf{r})d\mathbf{r} + \frac{1}{2}\int\int \frac{n(\mathbf{r})n(\mathbf{r}')}{|\mathbf{r}-\mathbf{r}'|}d\mathbf{r}d\mathbf{r}' \\
&+\int V_{ext}(\{\mathbf{R}_n\},\mathbf{r})n(\mathbf{r})d\mathbf{r} + \Omega_{xc}[n] + U(\{\mathbf{R}_n\}) \\
&+k_BT\sum_i\{f_i \ln f_i + (1-f_i)\ln(1-f_i)\},
\end{aligned} \tag{1}
$$

where $\Omega_{xc}[n]$ is an exchange-correlation functional at finite temperatures, $V_{ext}(\{\mathbf{R}_n\},\mathbf{r})$ an external potential due to ions at $\{\mathbf{R}_n\}$, $U(\{\mathbf{R}_n\})$ an ion-ion interaction potential, and the last term an entropy of noninteracting Fermi particles. The essential point here is that we now treat the occupation numbers $\{f_i\}$ as independent variables and introducing the chemical potential μ for the constancy of the particle number.

Once the stationary state which minimizes the grand potential is obtained, orbitals $\{\psi_i\}$ and occupation numbers $\{f_i\}$ satisfy the following equations:

$$
f_i\{H_{eff}-\mu\}\psi_i(\mathbf{r}) = \sum_j \lambda_{ij}\psi_j(\mathbf{r}) \quad , \tag{2}
$$

$$
f_i = \frac{1}{1+\exp\{[<\psi_i|H_{eff}|\psi_i>-\mu]/k_BT\}} \quad , \tag{3}
$$

where H_{eff} is the effective one particle Hamiltonian. We can prove [7] that (1) $<\psi_j|H_{eff}|\psi_i>=0$ if $f_i \neq f_j$, i.e. the effective Hamiltonian is block-diagonalized in a subspace of the same occupation number and the unitary transformation in this subspace leaves the grand potential invariant, and (2) the matrix $\{\langle\psi_j|H_{eff}|\psi_i\rangle\}$ is, not only block diagonal, diagonal with respect to i and j. Therefore, by treating at elevated temperatures the occupation numbers f_i as independent variables to minimize the the grand potential functionals, an additional diagonalization procedure is not necessary. Figure 5 shows an example of the bcc solid Na, where we use a unit cell of 16 atoms. The Fermi energy is located at a level ~0.03 (atomic unit). The degeneracy is 1, 6, 6, and 6 from bottom to up and the occupation number is 1, 1, 1/6 and 0, which satisfies the desired symmetry. This type of formalism may also be useful to a study of dynamics of the systems off the Born-Oppenheimer surface. [16]

5. Summary

We reviewed our recent work of the first-principle structural optimization by using the Car-Parrinello method. The Si(111)-($\sqrt{3}\times\sqrt{3}$)$R30°$-Al phase of 1/3 coverage was investigated. The Si(100)-Al interface structures with higher coverage show a new phase of metallic Al layers on Si substrate. A stable algorithm were proposed with using fractional occupations and finite temperatures, which provides a stable procedure to a goal without any diagonalization procedure and retains a symmetry required.

This work was supported by a Grant-in-Aid for Science Research from Japan Ministry of Education and Culture.

References

1. R.Car and M.Parrinello, Phys. Rev. Lett. **55**, 2471 (1985).
2. M.P.Teter, M.C.Payne, D.C.Allen, Phys. Rev.B**40**, 12255 (1989).
3. P.Pulay, Mol. Phys. **17**, 197 (1969).
4. D.Vanderbilt, Phys. Rev. B**41**, 7892 (1990).
5. J.M.Solar and A.R.Williams, Phys. Rev.B**40**, 1560 (1989): T.Oguchi, in this Proceedings.
6. H.Tsuge, M.Arai and T.Fujiwara, Jpn. J. Appl. Phys. **30**, L1583 (1991).
7. Y.Yamamoto and T.Fujiwara, to appear in Phys. Rev. in November 1992.
8. Y.Yamamoto and T.Fujiwara, un published work.
9. I. Yamada, H. Usui, S. Tanaka, U. Dahmen and K.H. Westmacott, J. Vac. Sci. Technol. a**8**(3), 1443 (1990); I. Yamada, Vacuum**41**, 889 (1990).

10. J. Nogami, A.A. Baski, and C.F. Quate, Phys. Rev. B**44**, 1415 (1991).
11. C.B.Bachelet, D.R.Hamann, and M.Schlüter, Phys. Rev. B**26**, 419 (1982).
12. D.R. Hamann, Phys. Rev B**40**, 2980 (1989).
13. D.M. Ceperley and B.J. Alder, Phys. Rev. B**18**, 3126 (1978).
14. T.Oguchi and T.Sasaki, Progr. Theor. Phys. Suppl. **103**, 93 (1991);
15. N.D.Mermin, Phys. Rev. **137**, A1441 (1965).
16. R.M.Martin, private communications.

Alkali-Metal Adsorption on the Si(001) Surface

Y. Morikawa[1], *K. Kobayashi*[2], *and K. Terakura*[1]

[1]Institute for Solid State Physics, University of Tokyo, Roppongi, Minato-ku, Tokyo 106, Japan

[2]National Institute for Research in Inorganic Materials, Namiki 1-1, Tsukuba-shi, Ibaraki 305, Japan

Abstract The alkali-metal (Li, Na, and K) adsorption on the Si(001) surface is studied by the first-principles molecular dynamics based on the norm-conserving pseudopotential. Our theoretical results agree well with various kinds of experiments and strongly support the double-layer model proposed by Abukawa and Kono for the Na and K saturated surfaces. It is shown that Li is very different from Na and K as adsorbates on the surface.

1. Introduction

The work function of metals and semiconductors decreases rapidly at the initial stage of alkali-metal adsorption and saturates, in many cases, after taking a shallow minimum at certain coverage [1]. On the other hand, oxygen adsorption generally tends to increase the work function. It is therefore very interesting that on the Cs saturated Si(001) surface, oxygen adsorption further reduces the work function leading to negative electron affinity (NEA).

In 1973, Levine [2] considered a mechanism of NEA and proposed a structural model (Levine's model). He assumed that alkali-metal atoms are adsorbed at the hollow site along the dimer rows (HH site, see Fig. 1) and oxygen atoms are at the cave site between the dimer rows (T4 site). He thought that the strong dipole layer which consists of positively charged alkali-metal atoms in the vacuum side and negatively charged oxygen atoms in the bulk side reduces the work function dramatically. In this model, alkali-metal atoms form one dimensional metallic chains along the dimer rows and the coverage of alkali-metal is $\Theta = 0.5$ with the coverage Θ defined as a ratio of the number of adatoms to that of the outer most substrate Si atoms. For many years, Levine's model for the Cs adsorption has been regarded as a proper model for the alkali-metal saturated Si(001) surface even for Na and K cases.

On the other hand, the angle resolved ultraviolet photoelectron spectroscopy (ARUPS) experiment [3] indicates that the K saturated Si(001)

Springer Series in Solid-State Sciences, Volume 114
Interatomic Potential and Structural Stability   Eds.: K. Terakura and H. Akai

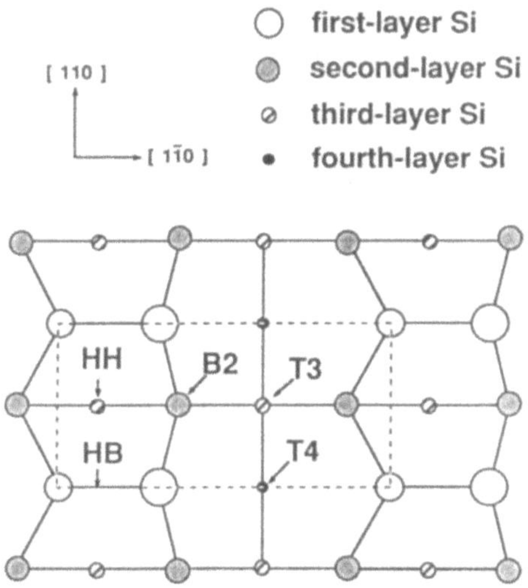

Fig. 1
A top view of the Si(001)-2 × 1 surface with asymmetric dimers. The 2 × 1 unit cell is indicated by dashed lines. The HH, HB, T3, T4, and B2 sites are indicated. The inter-dimer bridge site B2 is not necessarily just on top of the second layer Si atom.

surface is not metallic but semiconducting. Abukawa and Kono [4] proposed a new model called a double-layer (DL) model based on their x-ray photoelectron diffraction (XPD) experiment. In their model, alkali-metal atoms are adsorbed both on the HH and T3 sites and the coverage is $\Theta = 1.0$. A first-principles calculation by Ishida and Terakura [5] showed that the electronic structure of the DL model is actually semiconducting being consistent with the ARUPS result. Several experiments have been performed since then to check which model is correct. An impressive experiment by thermal desorption spectroscopy (TDS) [6] suggests that there should be two different K adsorption states with equal amount corresponding to the DL model. Besides, medium-energy-ion scattering [7], and reflection-high-energy-electron diffraction (RHEED) [8] also support the DL model. Although Levine's model is still supported by some experiments such as STM [9,10], LEED and AES [11-13], there are some ambiguities in the interpretation of these experimental data. The STM image [9,10] of the K saturated Si(001) surface clearly shows one dimensional bright chains along the dimer rows and their separation is just the same as that between two adjacent dimer rows. An intuitive interpretation is that the bright chains may correspond to alkali-metal atoms on the HH sites, thus supporting Levine's model. However, as indicated in the case of Si(111) $\sqrt{3} \times \sqrt{3}$-Ag [14,15], an intuitive interpretation sometimes fails. An STM image directly corresponds not to the atomic arrangement but to the local density of states around the Fermi level. We will demonstrate that the DL model is consistent with the STM image [9,10]. Glander and Webb [11,12] concluded from their LEED and AES experiment that the saturation coverage of Na adsorption on the Si(001) surface is $\Theta = 0.68$. They assumed the Na coverage of the well ordered 4×1 surface to be 0.25. Very recently, Abukawa and co-workers [16] studied Na adsorption on the single domain Si(001) surface by LEED and found that the LEED pattern is not 4×1 but $\begin{pmatrix} 4 & \pm 1 \\ 0 & 3 \end{pmatrix}$ in the matrix notation. Thus the coverage estimation of Glander and Webb is questionable. Wei *et al.* [13] claimed from their LEED I-V analysis at the saturation coverage that Levine's model can give a small R factor. They also found, however, that the R factor of the DL model is as small as that of Levine's model.

A first-principles calculation with full lattice relaxation for the alkali-metal adsorbed Si(001) surface was first performed by Batra [17]. He studied 0.5 and 1.0 ML alkali-metal adsorbed Si(001) surfaces using the norm-conserving pseudopotential [18] with the plane wave basis set. In his calculation, the bond length between the alkali-metal adatom and the surface Si atom is significantly underestimated due to the neglect of the non-linear contribution from the core charge density to the exchange-correlation energy. The partial core correction (PCC) developed by Louie *et al.* [19], is indispensable for the structural optimization associated with the alkali-metal atoms.

In our recent work [20-22], we also fully optimized atomic configuration of alkali-metal (Li, Na, and K) adsorbed Si(001) surfaces using first-principles molecular dynamics [23] based on the norm-conserving pseudopotential with the PCC. Our theoretical results strongly support the DL model and successfully reproduce various experimental results. In this article, we summarize our study.

2. Model and Calculational method

Our calculation is based on the local density approximation in the density functional theory [24,25] and the norm-conserving pseudopotential [18] with the plane wave basis set. We used a d potential as a local potential and s, p non-locality is taken into account. We constructed the pseudopotential of the occupied states (Si s, p and alkali-metal s) with the Bachelet-Hamman-Schlüter scheme, the pseudopotential of the unoccupied states (Si d, alkali-metal p and d) with the Hamman scheme [26] and transformed them to Kleinman and Bylander's separable type [27]. The PCC is incorporated for alkali-metals to improve the exchange-correlation energy and potential. The electronic degrees of freedom are optimized using the modified steepest descent method [28] with semi analytic integration [29]. The ionic degrees of freedom are optimized using the quenched molecular dynamics. The cut-off energy of the plane wave basis set is 6.25 Ry for the Na and K adsorbed systems and 12.25 Ry for the Li adsorbed system.

We used a repeated slab geometry and one slab consists of ten Si layers and a vacuum region of 13.6 Å thick. Alkali-metals are introduced on the both surfaces of each slab. Atoms in the two center layers are fixed at their ideal bulk position and other atoms are relaxed by the Hellmann-Feynman forces [30]. The convergence criterion for the forces acting on atoms is less than 3.0×10^{-3} Ry/a_H, with a_H denoting the Bohr radius (= 0.5292Å).

3. Results and Discussion

We optimized the alkali-metal adsorbed Si(001) surface with various coverages and estimated the coverage dependence of the adsorption energy defined by

$$E_{\mathbf{ad}}(\Theta_1 : \Theta_2) = E_{\mathbf{atom}} - \frac{E_{\mathrm{T}}(\Theta_2) - E_{\mathrm{T}}(\Theta_1)}{N_s(\Theta_2 - \Theta_1)},$$

where $E_{\mathrm{T}}(\Theta)$ is the total energy of the system with coverage Θ of alkali-metal adsorption and $E_{\mathbf{atom}}$ denotes the total energy of a free alkali-metal atom. $E_{\mathbf{atom}}$ is estimated by a total energy calculation for the bcc structure with a large lattice constant of 30.0 a.u. with the same cutoff energy of 6.25 Ry (12.25 Ry for Li), and the spin-polarization energy (−0.37 eV/atom for Li, −0.32 eV/atom for Na and −0.25 eV/atom for K) is separately estimated by an all electron free atom calculation. The results are shown in Fig. 2(a)–(c). For Na adsorption, the dashed line denotes the experimental result obtained by Glander and Webb [12]. Because we think that the saturation coverage should be 1.0 monolayer (ML) rather than 0.68 ML as discussed below, we rescaled their coverage estimation so that their saturation coverage may also be 1.0 ML. For K adsorption, dashed lines indicate the experimental desorption energy obtained by Tanaka *et al.* [6]. The agreement between theoretical and experimental results is satisfactory. The most important point is that for both Na and K adsorption, the adsorption energy below 1.0 ML is

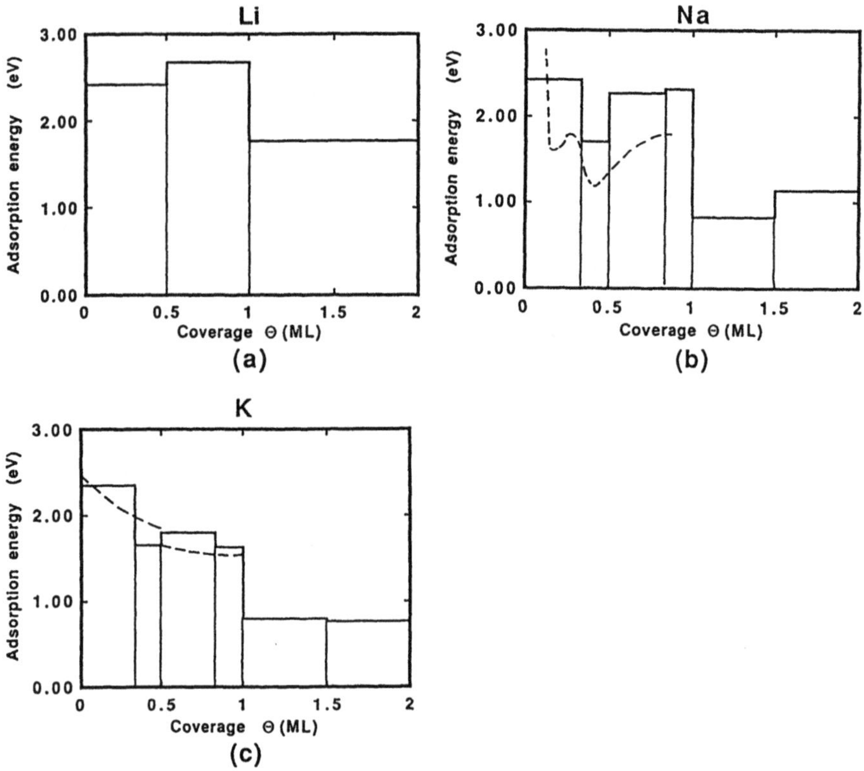

Fig. 2
The coverage dependence of the adsorption energy for (a) Li, (b) Na, and (c) K . These energies are obtained from the total energies for Θ = 0.0, 1/3, 0.5, 5/6, and 1.0, and a free Li (Na, K) atom. The Θ = 1.5 and 2.0 ML systems for both Na and K , and the 2.0 ML system for Li are also calculated to estimate the mean adsorption energy beyond 1.0 ML. The broken curves denote the experimental estimations by Glander and Webb [12] for Na adsorption and by Tanaka *et al.* [6] for K adsorption. Glander and Webb's results are plotted taking 1.0 ML as the saturation coverage.

larger than 1.6 eV and suddenly drops to less than 1.0 eV for higher coverages suggesting that 1.0 ML adsorption should be completed at room temperature. Experimentally it is observed that multilayer alkali-metals start to be desorbed at about 350 K [6,31]. Thus, Na and K atoms, deposited at around 350K, would not be adsorbed after completion of one monolayer. As for Li adsorption, the adsorption energy above 1.0 ML is 1.77 eV which is much larger than those of the Na and K cases and

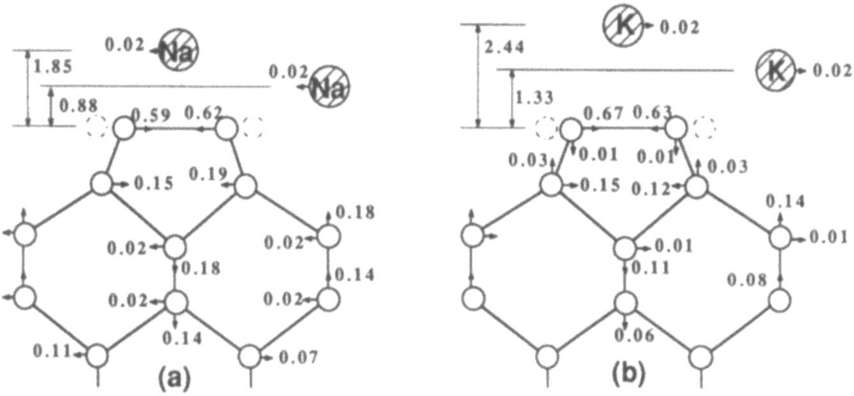

Fig. 3
The side views of the optimized structure for the HH-T3 site adsorption of (a) Na and (b) K. Displacements from the bulk terminated ideal surface are indicated in Å . The 1st layer Si atoms of the ideal surface are indicated by the dotted circles, with reference to which the height of Na (K) atoms (hatched circles) are measured. The horizontal displacements of Na (K) atoms are measured with reference to the 3rd layer Si atoms of the ideal surface.

slightly larger than the Li bulk cohesive energy of 1.63 eV. Therefore, Li would be adsorbed beyond 1.0 ML. We will discuss this point later.

The side views of the optimized structures of 1.0 ML Na and K adsorbed surfaces are shown in Fig. 3. Due to complete filling of the π^* state, the buckling of Si dimer is completely suppressed. The dimer bonds become longer than that of clean surface (2.26 Å) by 0.37 (0.28) Å for Na (K). It should be pointed out that the dimer bond length of the Na adsorbed surface is larger than that of the K adsorbed surface indicating stronger interaction between Na and the substrate than between K and the substrate which is consistent with the adsorption energy (see Fig. 2). These structures agree fairly well with experimental results. For example, the experimentally determined vertical separations between upper and lower alkali-metal atoms are 1.0 [16] and 1.1 Å [4] for Na and K respectively, while our theoretical results are 0.97 and 1.11 Å . Zhang *et al.* [32] also obtained a similar structure for Na adsorption using a similar method. Abukawa and co-workers [33,34] measured the band structures of Na and K saturated Si(001) surfaces using the synchrotron radiation ARUPS and compared their results with our results. They found that the agreement between theory and experiment is quite satisfactory for both Na and K adsorbed surfaces. Johansson and Reihl [35] measured the un-

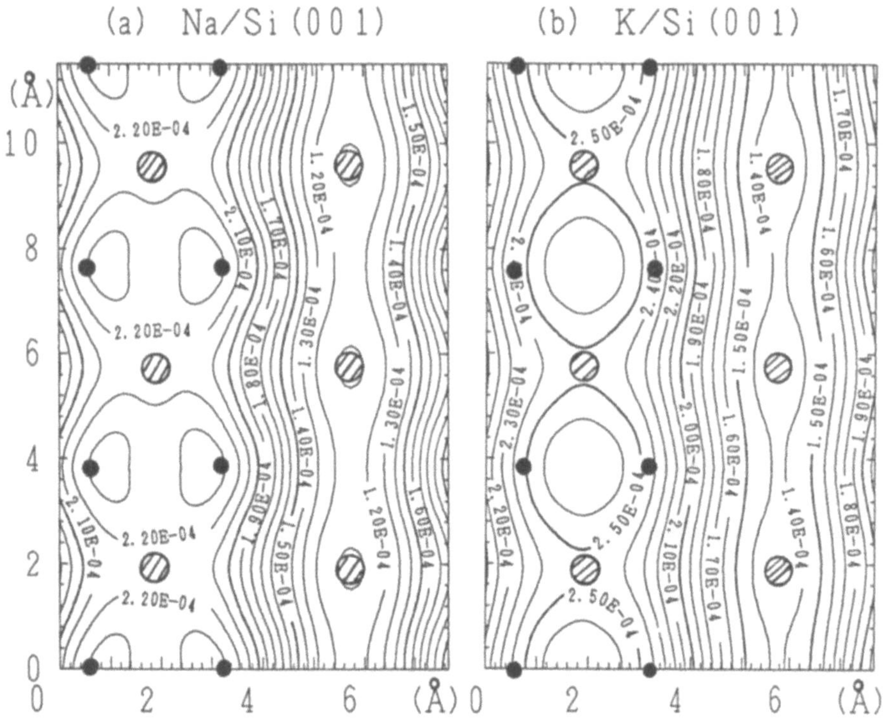

Fig. 4
Integrated local density of states maps for the occupied state for (a) Na and (b) K with the HH-T3 site adsorption within a plane parallel to the surface. The local density of states is integrated from 1 eV below the highest occupied level to the highest occupied level. The cut plane is located at 2.1 (2.6) Å above the Na (K) adatoms at the HH site. The difference between the adjacent contour lines is 10^{-5} a. u. The surface Si positions projected onto the plane are denoted by filled circles and the alkali-metal atom positions are by hatched circles.

occupied states of the K saturated surface by the inverse-photoemission spectroscopy (IPES). We showed in our previous paper [22] that the experimentally observed unoccupied metallic state with large dispersion is quantitatively well reproduced by our calculation. Very recently, they performed a similar experiment for the Na saturated surface [36] and found similar but less dispersive unoccupied states. Theoretical results do not have corresponding states in the gap region. In our calculation, these states may be pushed up into the bulk conduction band due to strong interaction between the Na atoms and the Si substrate.

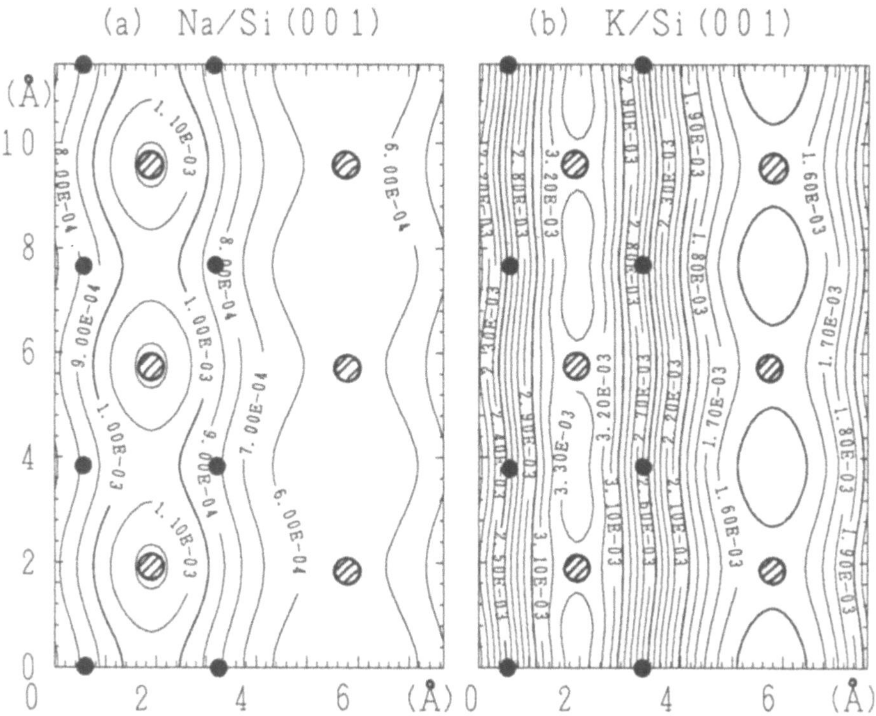

Fig. 5

Integrated local density of states maps for the unoccupied state for (a) Na and (b) K with the HH-T3 site adsorption within a plane parallel to the surface. The local density of states is integrated from the lowest unoccupied level to 1 eV above the lowest unoccupied level. The cut plane is located at 3.2 (2.0) Å above the Na (K) adatoms at the HH site. The difference between the adjacent contour lines is 10^{-4} a. u.

One controversial experimental data for the DL model is the STM image as discussed above [9,10]. In order to analyze the STM image, we calculated the integrated local density of states around the Fermi level in a plane parallel to the surface separated by 2.1 (2.6) Å from the Na (K) atoms along the dimer row. We integrated the density of states from 1 eV below the highest occupied level to the highest occupied level for the occupied state and from the lowest unoccupied state to 1 eV above the lowest unoccupied state for the unoccupied state. The integrated local density of states for the occupied and the unoccupied states are shown in Figs. 4 and 5 respectively. High density regions of the occupied states are located not above the alkali-metal atoms but above the Si-dimers, suggesting that both the lower and upper alkali-metal adatoms are completely invisible while only the dimer rows can be seen. Very recently, Motai *et al.* [37] obtained the STM image of Na saturated Si(001) surface. Their instrumental resolution is so high that they can clearly observe the separate two peaks along each dimer in Fig. 4(a). The unoccupied state is somewhat different from the occupied state. For Na adsorption, high density regions are located not above the dimers but above the Na adatoms. experimentally, this has not yet been observed. For K adsorption, the high density regions have weak corrugation along the dimer row indicating that this state is nearly free electron state which is consistent with large dispersion of unoccupied state as indicated in the IPES experiment [35] and our calculation. Although there are some differences between the occupied and unoccupied states, the lower layer alkali-metal atoms are invisible for both the occupied and unoccupied states. Therefore, the STM image [9,10,37] does not contradict the DL model.

Next, we discuss the low coverage region. Quite clear STM images for the very low coverage ($\sim$ 0.02 ML) Li, K and Cs adsorption were observed by Hasizume *et al.* [38]. They found that K and Cs atoms are adsorbed on the off-centered T3 site and Li atoms are on top of the upper Si atom of the buckled dimer and the buckling of Si dimer is stabilized near the alkali-metal atoms. Because it is impossible to perform calculations in such a low coverage region as 0.02 ML, we optimized the structure for the 1/12 ML K adsorbed surface instead by using a 4×3 unit cell, which enables us to simulate the local substrate structure around the K atom. The schematic top view of the (4×3)-K adsorbed surface is shown in Fig. 6(a). Fig. 6(b) shows the calculated local density of states integrated from 1 eV below the highest occupied level to the highest occupied level in the plane parallel to the surface separated by 2.6 Å from the K atoms along the trough. The amount of the buckling of the dimers adjacent to the K adatom is 0.63 Å which is slightly smaller than that for the clean surface by 0.08 Å . High density regions are located in between the K atoms and its nearest neighbor Si atoms. The enhancement of the local density of states around the adatom is consistent with the STM image. Although in the STM image, K atoms seem to be adsorbed at the off-centered T3 site, our calculated results indicate that K atom is adsorbed just at the T3 site. This discrepancy is however not so serious, because our calculation is for a much higher coverage compared to the experimental one.

Li behaves very differently from Na and K as adsorbates. According to the coverage dependence of the adsorption energy shown in Fig. 2, Li atoms continue to be adsorbed beyond 1.0 ML. The side views of the

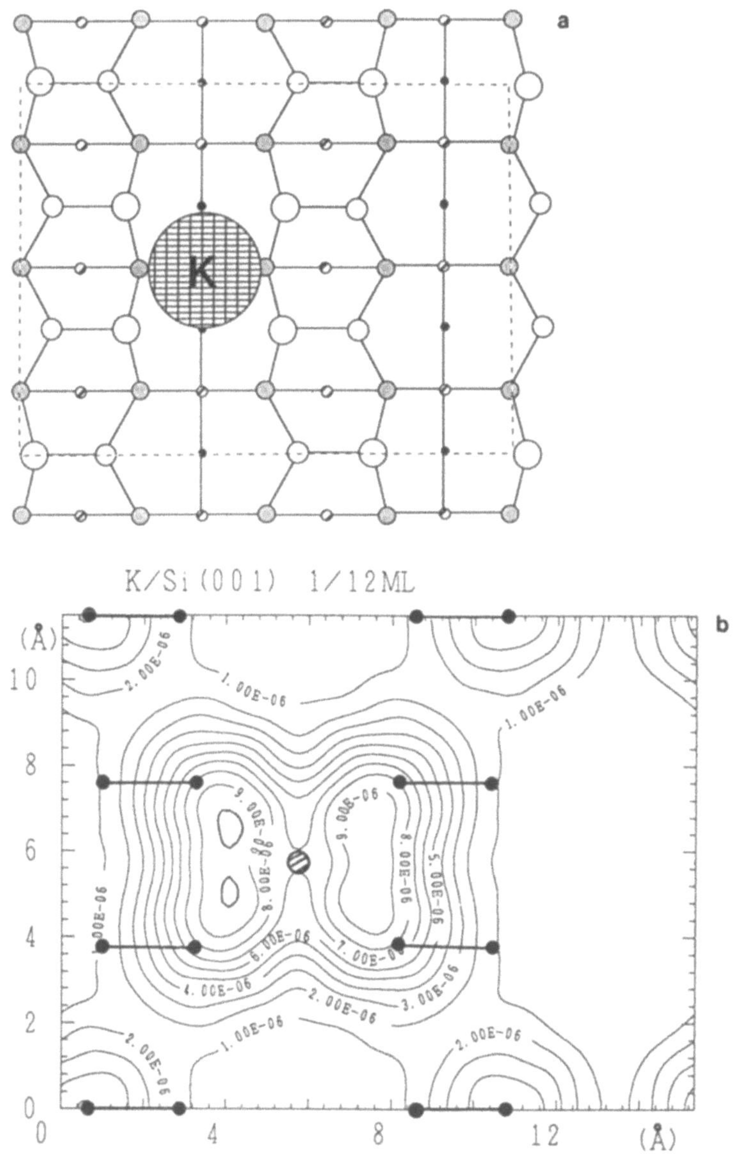

Fig. 6
(a) The schematic top view of the (4 × 3)-K adsorbed surface. (b) Integrated local density of states maps for the occupied state for 1/12 ML K adsorption within a plane parallel to the surface. The local density of states is integrated from 1 eV below the highest occupied level to the highest occupied level. The cut plane is located at 2.6 Å above the K adatom. The difference between the adjacent contour lines is 10^{-6} a. u.

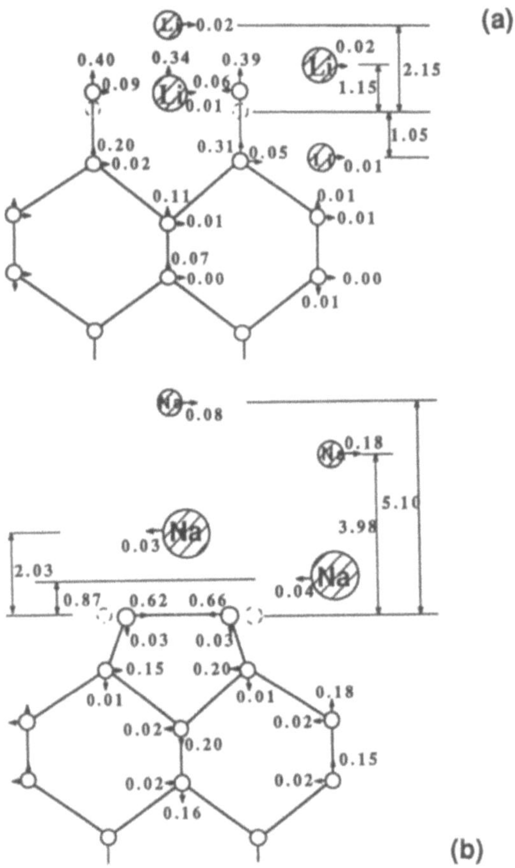

Fig. 7
The side views of the optimized structure for the 2.0 ML adsorption of (a) Li and (b) Na. Large hatched circles indicate alkali-metal atoms adsorbed at the HH or T3 sites and small hatched circles at the HB or T4 sites.

2.0 ML Li and Na adsorbed structures are shown in Fig. 7(a) and (b), respectively. For 2.0 ML Li adsorption, the lower layer Li atoms go deep into the substrate up to almost the same level as the second layer Si atoms and the first layer Si dimerization is completely broken. This is consistent with the experimentally observed 1×1 LEED pattern [39]. For the Na case, the first layer Si dimerization remains and the dimer bond length is even shorter than that of the 1.0 ML Na adsorption case. This difference between Li and Na may come from the atomic size difference. The upper layer Na atoms reduce the bond strength between the lower layer Na atom and first-layer Si atom, and thereby the dimer bond strength is enhanced.

On the other hand, the upper layer Li atoms can directly interact with the first-layer Si atoms and break the first-layer Si dimerization. It is also important to note that the Si-Si bond length between the first and second layers is significantly elongated. Therefore, the first-layer Si atoms may be easily detached from the rest bulk Si. Then the Si atoms originally in the second layer form the topmost Si surface and Li adatoms can move further into the bulk Si. We believe that such a process may be the initial stage of the Li-silicide formation.

4. Conclusion

We studied the alkali-metal (Li, Na and K) adsorbed Si(001) surfaces by using the first-principles molecular dynamics method. Our theoretical results strongly support the double layer model [4] corresponding to 1 ML for the Na and K saturated surfaces and can reproduce various experimental data. We demonstrated that Li behaves very differently from Na and K on the Si(001) surface. Li atoms continue to be adsorbed beyond 1.0 ML and, they break the first layer Si dimerization completely at 2.0 ML. This is consistent with the experimental LEED 1×1 pattern and may be the initial stage of Li-silicides formation. At low coverage, K atom is adsorbed at the T3 site and surrounded by four raised Si atoms of the neighboring buckled dimers. The occupied part of the local density of states within 1 eV from the Fermi level is peaked in between the K atom and its neighboring Si atoms.

Acknowledgement

The authors are grateful to Dr. H. Ishida, Professor S. Kono, Professor T. Sakurai, Professor T. Hashizume and their co-workers for valuable discussions. They also thank Dr. S. Blügel for the collaboration in the early stage of this study. The numerical calculations were performed at the computer centers of University of Tokyo, Hokkaido University, Institute for Solid State Physics, and Institute for Molecular Science. This work was partially supported by a Grant-in-Aid for Scientific Research on Priority Areas from Ministry of Education, Science, and Culture.

References

1 For general references, see articles in *Physics and Chemistry of Alkali Metal Adsorption,* edited by H. P. Bonzel, A. M. Bradshaw, and G. Ertl, (Elsevier, Amsterdam, 1989).
2 J. D. Levine, Surf. Sci. **34**, 90 (1973).
3 Y. Enta, T. Kinoshita, S. Suzuki, and S. Kono, Phys. Rev. B **36**, 9801 (1987).
4 T. Abukawa and S. Kono, Phys. Rev. B **37**, 9097 (1988).
5 H. Ishida and K. Terakura, Phys. Rev. B **40**, 11519 (1989).
6 S. Tanaka, N. Takagi, N. Minami, and M. Nishijima, Phys. Rev. B **42**, 1868 (1990).
7 A. J. Smith, W. R. Graham, and E. W. Plummer, Surf. Sci. **243**, L37 (1991).
8 T. Makita, S. Kohmoto, and A. Ichimiya, Surf. Sci. **242**, 65 (1991).
9 P. Soukiassian and J. A. Kubby, in *The Structure of Surfaces III,* edited by S. Y. Tong, M. A. Van Hove, K. Takayanagi, and X. D. Xie, (Springer-Verlag, Berlin, 1991), p. 584.
10 J. A. Kubby, W. J. Greene, and P. Soukiassian, J. Vac. Sci. Technol. **B9** 739 (1991).
11 G. S. Glander and M. B. Webb, Surf. Sci. **222**, 64 (1989).
12 G. S. Glander and M. B. Webb, Surf. Sci. **224**, 60 (1989).
13 C. M. Wei, H. Huang, S. Y. Tong, G. S. Glander, and M. B. Webb, Phys. Rev. B **42** 11284 (1990).
14 Y. G. Ding, C. T. Chan, and K. M. Ho, Phys. Rev. Lett. **67**, 1454, (1991).
15 S. Watanabe, M. Aono, and M. Tsukada, Phys. Rev. B **44** 8330 (1991).
16 T. Abukawa, H. Takahashi, and S. Kono, the 46th meeting of the Phys. Soc. of Japan 30a ZD-12 (Sapporo, Japan, Sep. 1991).
17 I. P. Batra, J. Vac. Sci. Technol. **A8**, 3425 (1990).
18 G. B. Bachelet, D. R. Hamann, and M. Schlüter, Phys. Rev. B **26**, 4199 (1982).
19 S. G. Louie, S. Froyen, and M. L. Cohen, Phys. Rev. B **26**, 1738 (1982).
20 Y. Morikawa, K. Kobayashi, K. Terakura, and S. Blügel, Phys. Rev. B **44**, 3459 (1991).
21 K. Kobayashi, Y. Morikawa, K. Terakura, and S. Blügel, Phys. Rev. B **45** 3469 (1992).
22 Y. Morikawa, K. Kobayashi, and K. Terakura, Surf. Sci. (to be published).
23 R. Car and M. Parrinello, Phys. Rev. Lett. **55**, 2471 (1985).

24 P. Hohenberg and W. Kohn, Phys. Rev. **136**, B864 (1964).
25 W. Kohn and L. J. Sham, Phys. Rev. **140**, A1133 (1965).
26 D. R. Hamann, Phys. Rev. B**40**, 2980 (1989).
27 L. Kleinman and D. M. Bylander, Phys. Rev. Lett. **48**, 1425 (1982).
28 A. R. Williams and J. Soler, Bull. Am. Phys. Soc. **32**, 562 (1987).
29 M. C. Payne, J. D. Joannopoulous, D. C. Allen, M. P. Teter, and D. H. Vanderbilt, Phys. Rev. Lett. **56**, 2656 (1986).
30 For the Na case, atoms in the two center layers are allowed to be relaxed. As indicated in Fig. 3(a), the displacement is only 0.1 Åwhich is so small that it would not affect any results of this study.
31 M. Tikhov, G. Boishin, and L. Surnev, Surf. Sci. **241**, 103 (1991).
32 B. L. Zhang, C. T. Chan, and K. M. Ho, Phys. Rev. B **44** 8210 (1991).
33 T. Abukawa, T. Kashiwakura, T. Okane, Y. Sasaki, H. Takahashi, Y. Enta, S. Suzuki, S. Kono, S. Sato, T. Kinoshita, A. Kakizaki, T. Ishii, C. Y. Park, S. W. Yu, K. Sakamoto, and T. Sakamoto, Surf. Sci. **261**, 217 (1992).
34 T. Abukawa, T. Kinoshita, C. Y. Park, T. Kashiwakura, T. Okane, Y. Sasaki, H. Takahashi, S. W. Yu, K. A. Kang, Y. Enta, S. Suzuki, S. Kono, S. Sato, A. Kakizaki, T. Ishii, K. Sakamoto, and T. Sakamoto, the 47th meeting of the Phys. Soc. of Japan, 27p ZS-9 (Yokohama, Japan, Mar. 1992).
35 L. S. O. Johansson and B. Reihl, Phys. Rev. Lett. **67** 2191 (1991).
36 L. S. O. Johansson and B. Reihl, (to be published).
37 K. Motai, T. Hashizume, H. Lu, D. R. Jeon, and T. Sakurai, the 47th meeting of the Phys. Soc. of Japan, 30a ZF-12 (Yokohama, Japan, Mar. 1992).
38 T. Hashizume, I. Sumita, Y. Murata, S. Hyodo, and T. Sakurai, J. Vac. Sci. Technol. **B9**, 742 (1991).
39 H. Tochihara and Y. Murata, Surf. Sci. **215** L323 (1989).

Structural and Vibrational Properties of Alkali Overlayers on Metals

H. Ishida and Y. Morikawa

Institute for Solid State Physics, University of Tokyo,
Roppongi, Minato-ku, Tokyo 106, Japan

Abstract. Structual and vibrational properties of Na overlayers on the semi-infinite jellium with r_s=2 are studied as a function of coverage (θ) within local-density-functional theory. The weakening in the adatom–substrate (A-S) bond with increasing θ leads to the outward relaxation of the overlayer and the decrease in the A-S stretching frequency. The dynamical charge associated with the A-S stretching mode decreases with increasing θ much more rapidly than the static adatom dipole moment, which is in agreement with the recent electron energy loss experiments.

I. Introduction

Within the second-moment approximation, the strength of a nearest-neighbour (NN) bond in metals scales as $N^{-1/2}$ as a function of the NN coordination number N. [1] In case of chemisorption, this simple statement on many-atom interaction implies that the adsorbate–substrate (A-S) bond is weakened with the increasing adsorbate–adsorbate (A-A) interaction. One of the best candidates to observe this effect may be adsorption of alkali metals, since extended valence s orbitals of nearby alkali atoms overlap significantly in a wide coverage (θ) range. The classical Gurney model for these adsorbates, on the other hand, tried to explain the θ dependence of the adsorbate electronic structure only in terms of the indirect dipole–dipole interaction between adatom dipole moments.[2] However, recent density-functional calculations have demonstrated that the direct A-A interaction due to the orbital overlap plays a more important role except at very low θ.[3 – 5]

Alkali-metal adsorption has been studied extensively as a prototype chemisorption system exhibiting interesting properties such as large work-function changes and catalytic promotion.[6] Over the last few years, there was significant progress in the study of these adsorbates. On the theoretical side, detailed knowledge of the adsorbate electronic structure was obtained from density-functional calculations as mentined above. On the experimental side, there has been growing

Springer Series in Solid-State Sciences, Volume 114
Interatomic Potential and Structural Stability Eds.: K. Terakura and H. Akai

interest in the study of structual and vibrational properties of the overlayer. Among them are the adsorption geometry, the A-S bond length, and the A-S stretching mode.[7 – 14] New findings that emerged from these experiments have not completely been understood because of the lack of the theory relevant to them. In the present article, we discuss the θ dependence of the structual and vibrational properties of alkali overlayers based on our density-functional calculation. The Hartree atomic units are used throughout the text.

II. Model and Method

To simulate adsorption on high-density metals, we utilize the semi-infinite jellium as substrate. When the substrate atoms are replaced by a flat jellium surface, the substitutional adsorption and vibrations in the plane cannot be discussed. On the other hand, one can concentrate on a fundamental question as to how the nature of the A-S bond changes with increasing θ, since the ratio of the strength of the A-A interaction over that of the A-S interaction is the only parameter governing the present system.

Our calculation is performed within the local density approximation to density-functional theory.[15] The one-electron Green function of the semi-infinite system is calculated self-consistently with use of the embedding metod of Inglesfield.[16] Alkali ion cores are represented by a Kleinman-Bylander pseudopotential[17] that incorporates the partial core correction of Louie *et al.*[18] The overlayer is assumed to form a hexagonal lattice, and θ is controlled by changing the NN A-A bond length $a_{\parallel}$. The only parameter in the present model is the distance between the Na plane and the edge of the background density of jellium, z_a. For each θ, calculations are performed for several values of z_a to determine the potential energy curve. More details of the computational method is found in Ref. 5.

III. Results and Discussion

We present results for Na overlayers on jellium with r_s=2, which approximately corresponds to the electron density of Al. For the sake of convenience, the coverage corresponding to $a_{\parallel}$=7.18 a.u. is defined as θ=1 in the following. Fig. 1 shows contour maps of the electron density on the vertical-cut plane containing NN Na atoms. The middle panel corresponds to a coverage near the work-function minimum. At lower θ, the outermost contours are highly corrugated and atomiclike, while

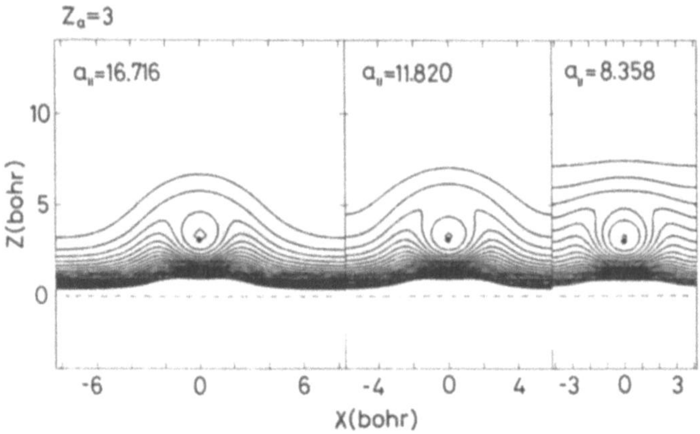

Fig.1 Contour maps of the electron density on a vertical-cut plane containing NN Na atoms for three Na overlayers on jellium. The jellium edge is located at z=0, and Na nuclei are indicated by solid circles. z_a is chosen as 3 a.u. for all θ.

the corresponding ones at higher θ are mostly parallel to the surface, which indicates formation of a quasi 2-dimensional metallic overlayer due to the large orbital overlap.

We expand the total energy of the system per Na adatom as

$$E_t(z_a) = E_0 + \frac{E_2}{2}(z_a - z_0)^2 + \frac{E_3}{6}(z_a - z_0)^3 + \cdots, \qquad (1)$$

where z_0 denotes the equilibrium position of the overlayer, and E_0 is chosen such that $E_t(z_a)$ tends to zero as $z_a \to \infty$, i.e., E_0 gives the binding energy of the overlayer.

The calculated z_0, E_0, E_2, and E_3 for several θ values are listed in Tab. 1. E_0 decreases monotonously by $\sim$0.7 eV in the θ range studied, which implies that the A-S bond is significantly weakened with increasing θ. Qualitatively speaking, this bond weakening takes place because Na valence electrons contribute to the A-A bonding in addition to the A-S bonding at higher θ as mentioned in Introduction. Following the weakening in the A-S bond, the A-S bond length is prolonged, and the curvature and anharmonicity of the potential energy curve become smaller with increasing θ. Lang and Williams[19] obtained z_0=3.1 a.u. as the bond length in the low θ limit. The increase in z_0 relative to their value amounts to $\sim$0.6 a.u. at θ=1 where the lattice constant $a_{\parallel}$ is close to the NN bond length in bulk Na.

The last two rows of Tab. 1 show the stretching frequency of the A-S bond in the harmonic approximation, $\hbar\omega_e=[E_2/M]^{1/2}$ and

Table 1. Calculated structual and vibrational parameters for Na overlayers on jellium with r_s=2. The lattice constant $a_{\parallel}$ at θ=1 is 7.18 a.u.

θ	0.185	0.369	0.553	0.738	1.00
z_0(a.u.)	3.26	3.33	3.43	3.59	3.69
$-E_0$(eV)	1.20	0.90	0.74	0.57	0.47
$E_2(10^{-2}$a.u.)	1.80	1.56	1.31	1.08	0.90
$-E_3(10^{-2}$a.u.)	2.79	2.18	2.01	1.59	1.31
$\hbar\omega_e$(meV)	17.8	16.6	15.2	13.8	12.6
$\sqrt{\langle(z_a - z_0)^2\rangle}$(a.u.)	0.135	0.140	0.146	0.153	0.160

the root mean square displacement in the ground state, $\langle(z_a - z_0)^2\rangle^{1/2}=[\hbar/(2M\omega_e)]^{1/2}$, where M denotes the atomic mass of Na. $\hbar\omega_e$ corresponds to a vibrational mode where all the Na atoms oscillate in phase relative to the substrate (2-dimensional wave vector $\mathbf{Q}$=0). The stretching frequency at the lowest θ, ~18 meV is in fair agreement with the measured frequency for Na/Cu, ~20 meV.[14,15] If Eq. (1) is fitted to the Morse potential $E_t(z_a) = D_e[1 - \exp(-a(z_a - z_0))]^2$ around z_0, the vibrational energies $\hbar\omega_n$=$\hbar\omega_e(n + 1/2)$ are replaced by $\hbar\omega_e(n + 1/2)[1 - x_e(n + 1/2)]$, where $x_e = \hbar\omega_e/(4D_e)$ provides a dimensionless measure of the anharmonicity.[20] From E_3, we obtain x_e ~0.005 at θ=0.18 and even smaller values at higher θ. Thus the anharmonicity effects may be small for the present system.

Astaldi *et al.*[13] and Lindgren *et al.*[14] found in their recent electron energy loss experiments for Na/Cu that the intensity of the A-S stretching mode nearly vanishes near the work-function minimum, which indicates that the dynamical charge associated with the A-S stretching mode decreases with increasing θ much more rapidly than the static dipole moment. We expand the adatom dipole moment per Na atom around z_0 as

$$p(z_a) = p_0 + p_1(z_a - z_0) + \cdots,$$

where p_0 and p_1 denote the static dipole monent at z_0 and the dynamical charge, respectively. The work-function change is given by $\Delta\Phi = 4\pi\theta p_0/S$, where S is the unit area at θ=1. Fig. 2(a) and (b) show the calculated p_0 and p_1, respectively. The solid circles at θ=0 indicate the data given by Lang and Williams[19] in the low θ limit. The dynamical charge p_1 actually decreases with increasing θ much more rapidly than the static dipole moment p_0 at low θ and changes its sign

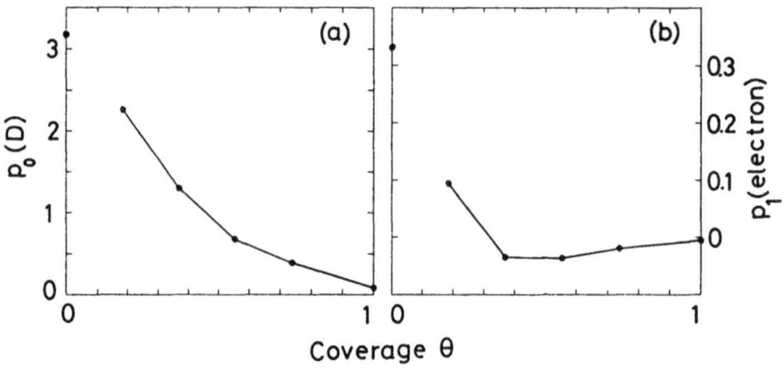

Fig.2 (a) Static adatom dipole moment p_0 and (b) Dynamical charge p_1 for hexagonal Na overlayers on jellium with r_s=2 as a function of θ. The solid circles at θ=0 indicate the data in Ref. 19.

near the work function minimum. It may be difficult to observe the A-S stretching mode at higher θ as the dipole scattering probability for an incoming electron is proportional to p_1^2.

To understand the θ dependence of the dynamical charge p_1, we first calculate the adatom density of states $\rho(\epsilon)$ and the dipole density of states $\mu(\epsilon)$ defined, respectively, by,

$$\rho(\epsilon) = \frac{-1}{\pi} \int_R d\mathbf{r} \mathrm{Im}\Big[G(\mathbf{r},\mathbf{r},\epsilon,\theta) - G(\mathbf{r},\mathbf{r},\epsilon,\theta=0)\Big],$$

$$\mu(\epsilon) = \frac{-1}{\pi} \int_R d\mathbf{r}(z_a - z) \mathrm{Im}\Big[G(\mathbf{r},\mathbf{r},\epsilon,\theta) - G(\mathbf{r},\mathbf{r},\epsilon,\theta=0)\Big],$$

where $G(\mathbf{r},\mathbf{r},\epsilon,\theta)$ denotes the Green function at the one-electron energy ϵ, and the volume integration is performed in a sphere with radius R centered at a Na atom. The dipole moment induced in the sphere is obtained by integrating $\mu(\epsilon)$ up to the Fermi energy E_F. In Fig. 3, we show the calculated $\rho(\epsilon)$ and $\mu(\epsilon)$ for three different values of z_a at θ=0.18 where the orbital overlap between nearby Na atoms is very small. The sharp atomic resonance above E_F is an antibonding state whose wave function is strongly polarized toward the vacuum, whereas the occupied bonding states are polarized toward the interface. Irrespective of the value of z_a, the dipole density of states $\mu(\epsilon)$ changes its sign very rapidly near E_F. The densities of states in Fig. 3 become sharper with increasing z_a, since the orbital overlap between the adatom and substrate states is reduced. At the same time, the energy separation between the broad maximum in $\mu(\epsilon)$ in the occupied part and the sharp minimum in the unoccupied part becomes smaller,

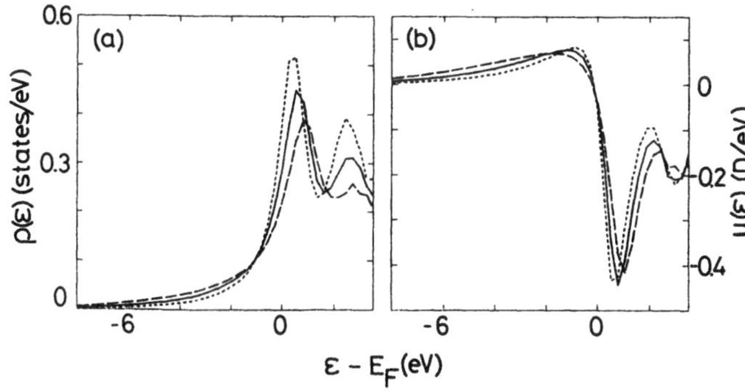

Fig.3 (a) Adatom density of states $\rho(\epsilon)$ and (b) dipole density of states $\mu(\epsilon)$ for Na overlayers on jellium with $r_s = 2$ at θ=0.18 ($a_{\|}$=16.72 a.u.). The sphere radius R is chosen as 3.5 a.u. The solid, dashed, and dotted curves correspond to z_a=3.25, 2.75, and 3.75 a.u. respectively.

which may be interpreted as a smaller bonding–antibonding (B-A) energy splitting.

To discuss the θ dependence of the dynamical charge, we consider two Na atoms on jellium for the four cases: $a_{\|}=a_1$ and a_2 ($a_1 < a_2$), and $z_a=z_1$ and z_2 ($z_1 < z_2$). a_1 and a_2 simulate the high and low θ regions, respectively. When the two atoms are well separated ($a_{\|}=a_2$), the B-A boundary in $\mu(\epsilon)$ for each atomic resonance ϕ_i is close to E_F as seen from Fig. 3, so that each atom has the largest dipole moment by the maximum use of the bonding states polarized toward the interface. Fig. 4(a) shows schematically $\mu(\epsilon)$ of each atomic resonance ϕ_i (i=1,2) for $z_a=z_1$ (dashed line) and for $z_a=z_2$ (solid line). $\mu(\epsilon)$ for $z_a=z_2$ is sharper than that for $z_a=z_1$ as discussed above. The positive dynamical charge at low θ implies $p(z_2) > p(z_1)$ at $a_{\|}=a_2$.

With the decreasing A-A distance ($a_{\|}=a_1$), the two atomic resonances split into the bonding molecular resonance $[m_1=(\phi_1+\phi_2)/\sqrt{2}]$ and the antibonding molecular resonance $[m_2=(\phi_1-\phi_2)/\sqrt{2}]$. In this case, as shown in Fig. 4(b), part of the strong antibonding resonance polarized toward the vacuum is occupied for m_1, while part of the bonding states is unoccupied for m_2. These two effects work cooperatively to decrease $p(z_a)$ with decreasing $a_{\|}$. Here it is important that these effects work more drastically for $z_a=z_2$ than for $z_a=z_1$, since larger part of the antibonding (bonding) states becomes occupied (unoccupied) as for m_1 (m_2) for a larger value of z_a because the resonance becomes sharper with increasing z_a. Thus $p(z_2)$ decreases with decreasing $a_{\|}$ more rapidly than $p(z_1)$. As a result, $p(z_2)$ can be

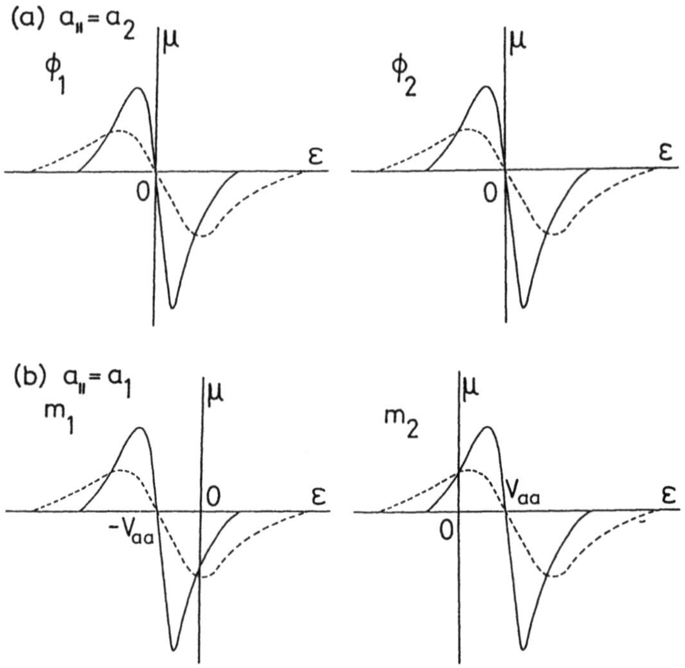

Fig.4 (a) Dipole density of states $\mu(\epsilon)$ for atomic resonances ϕ_i (i=1, 2) at a large A-A distance ($a_\parallel$=a_2). Solid and dashed curves correspond to z_a=z_2 and z_1 ($z_1 < z_2$), respectively. (b) Corresponding $\mu(\epsilon)$ for molecular resonances m_i (i=1, 2) at a smaller A-A distance ($a_\parallel$=a_1). a_1 and a_2 ($a_1 < a_2$) simulate high and low θ regions, respectively.

smaller than $p(z_1)$ at some smaller $a_\parallel$, which means that the dynamical charge changes negative at this θ. It is expected that this mechanism takes place near the work-function minimum in the present system, since the width of the overlayer bands already amounts to $\sim$ 1 eV in that θ range.

We believe that the above-mentioned mechanism may be applicable for systems where the B-A boundary with respect to the A-S bonding is located close to E_F at low θ. However, it will depend on the detailed shape of densities of states whether the dynamical charge really changes its sign at some θ.

IV. Conclusion

We have studied structual and vibrational properties of alkali overlayers on a metal substrate within density-functional theory. The A-S bond becomes weaker with the increasing A-A orbital overlap, which may be viewed as a consequence of the many-atom interaction. The weakening in the A-S bond leads to the outward relaxation of the overlayer and the lowering in the A-S stretching frequency. We proposed a new mechanism to explain the experimentally observed rapid decrease of the dynamical charge associated with the A-S stretching mode.

References

1. *Many-Atom Interactions in Solids*, edited by R.M. Nieminen, M.J. Puska, and M.J. Manninen (Springer, Berlin, 1990).
2. R.W. Gurney, Phys. Rev. **47**, 479 (1935).
3. E. Wimmer, A.J. Freeman, J.R. Hiskes, and A.M. Karo, Phys. Rev. B**28**, 3074 (1983).
4. H. Ishida and K. Terakura, Phys. Rev. B**38**, 5752 (1988).
5. H. Ishida, Phys. Rev. B**42**, 10899 (1990); Surf. Sci. **242**, 341 (1991).
6. *Physics and Chemistry of Alkali Adsorption*, edited by H.P. Bonzel, A.M. Bradshaw, and G. Ertl (Elsevier, Amsterdam, 1989).
7. D. Fisher, S. Chandavarkar, I.R. Collins, R.D. Diehl, P. Kaukasonia, and M. Lindroos, Phys. Rev. Lett. **68**, 2786 (1992).
8. H. Over, H. Bludau, M. Skotte-Klein, G. Ertl, W. Moritz, C.T. Campbell, Phys. Rev. B**45**, 8638 (1992).
9. A. Schmalz, S. Aminpirooz, L. Becker, J. Haase, J. Neugebauer, M. Scheffler, D.R. Batchelor, D.L. Adams, and E. Bogh, Phys. Rev. Lett. **67**, 2163 (1991).
10. J.N. Andersen, M. Qvarford, R. Nyholm, J.F. van Acker, E. Lundgren, Phys. Rev. Lett. **68**, 94 (1992).
11. M. Kerkar, D. Fisher, D.P. Woodruff, R.G. Jones, R.D. Diehl, and B. Cowie, Phys. Rev. Lett. **68**, 3204 (1992).
12. G.M. Lamble, R.S. Brooks, D.A. Kings, and D. Norman, Phys. Rev. Lett. **61**, 1112 (1988).
13. C. Astaldi, P. Rudolf, and S. Modesti, Solid State Commun. **75**, 847 (1990).
14. S.-A. Lindgren, C. Svensson, L. Wallden, Phys. Rev. B**42**, 1467 (1990).
15. W. Kohn and L.J. Sham, Phys. Rev. **140**, A1133 (1965).
16. J.E. Inglesfield, J. Phys. C **14**, 3795 (1981).

17. L. Kleinman and D.M. Bylander, Phys. Rev. Lett. **48**, 1425 (1982).
18. S.G. Louie, S. Froyen, and M.L. Cohen, Phys. Rev. B**26**, 1738 (1982).
19. N.D. lang and A.R. Williams, Phys. Rev. B**18**, 615 (1978).
20. See, for example, *Electron Energy Loss Spectroscopy and Surface Vibrations*, by H. Ibach and D.L. Mills (Academic, 1982, New York).

Atomic and Electronic Structure of Some Exotic Surfaces Revealed by Scanning Tunneling Microscopy

M. Tsukada

Department of Physics, Faculty of Science, University of Tokyo, Hongo 7-3-1, Bunkyo-ku, Tokyo 113, Japan

Abstract. A first-principles theoretical method of simulation for scanning tunneling microscopy image and spectroscopy data is presented based on the local density functional approach. The theoretical simulation plays very essential role to decode all the rich and detailed information from experiments, and to clarify effects of the microscopic structure of the tip. The method is applied to some interesting surface systems such as graphite, Si(100) dimer surfaces and step structures, Si(111)$\sqrt{3}\times\sqrt{3}$ - Ag surface. The mechanism to achieve the atomic resolution by the tip with the curvature of several hundreds of Angstroms is clarified and it is discussed how the atomic structure of the tip influences the STM image. Case studies show that naive interpretation of the STM image sometimes fails. Exotic phenomena such as transparency of the adsorbed molecule and a moire pattern formation are discussed.

1.Introduction

Recent remarkable progress of scanning tunneling microscopy (STM) made it possible to observe various exotic surface structures in the direct space with atomic resolution.[1] Systems without any translational periodicity, e.g., extended defects or intermediate states of surface chemical reactions can be studied in a microscopic way for the first time. Moreover, the scanning tunneling spectroscopy (STS) provides information of surface electronic structure at specified surface positions.[2] Therefore STM/STS can afford rich detailed information of microscopic surfaces. However it is rather difficult to decode such profound information from the experimental data, because the tip and the surface interact directly each other. To deconvolute the tip effects is far from trivial and needs theoretical works. Further so many kinds of mysterious phenomena have been reported which are closely related to the microscopic tunneling mechanism in the STM process. To make a firm basis for these phenomena is also an important target of the theoretical study.

So far we have developed the first-principles theory of simulation of STM images and STS spectra.[3] In our theory the real atomic structure of the tip is taken into account and its significant effect in achieving the atomic resolution can be clarified. In this article, we briefly introduce the method of the theoretical simulation and report some typical results of the simulation to reveal the effect of the tip. The systems we choose are graphite[4], Si(100) dimer surface[5] and the step structure[6], and the Si(111)$\sqrt{3}\times\sqrt{3}$ -Ag surfaces[7,8,9]. As an unusual feature of the tunneling we focus on the transparent images of weakly chemisorbed molecules and a moire pattern formation[10].

2. Formalism of the Theoretical Simulation and the Mechanism of STM

A straightforward extension of Bardeen's perturbation theory[11] to the STM system leads to the following expression of the tunnel current[12],

$$I = \frac{2\pi e}{\hbar}\int dE[f(E) - f(E-eV)]A(\mathbf{R}, E, E-eV) \tag{1}$$

$$A(\mathbf{R}, E, E') = \int d\mathbf{r}\int d\mathbf{r}' V_T(\mathbf{r})V_T(\mathbf{r}')\mathcal{G}^T(\mathbf{r},\mathbf{r}';E')\mathcal{G}^S(\mathbf{r}+\mathbf{R},\mathbf{r}'+\mathbf{R};E)\,. \tag{2}$$

Springer Series in Solid-State Sciences, Volume 114
Interatomic Potential and Structural Stability Eds.: K. Terakura and H. Akai

In the above $\mathcal{G}^T$ and $\mathcal{G}^S$ are the imaginary part of the Green's function for the tip and the surface, respectively, f is the Fermi distribution function, V_T is the potential of the tip, and V is the tip bias. Since $\mathcal{G}^S$ is a rapidly decreasing function in the direction perpendicular to the surface, the integral of eq.(2) is almost determined by a very narrow spatial region around the protrusion on the tip closest to the surface. The expansion of the smooth part of the surface Green's function around the center of mass of the weight function made up of the product of the tip wavefunction, tip potential, and the surface decay factor is efficient and leads to a moment expansion for the tunnel conductance[4]:

$$\frac{dI}{dV} = \frac{2\pi e^2}{\hbar} A(\mathbf{R}, E_F + eV, E_F) - \frac{2\pi e^2}{\hbar} \int_{E_F}^{E_F+eV} dE \frac{\partial A(\mathbf{R}, E', E - eV)}{\partial E} |_{E'=E} \quad (3)$$

Neglecting the second term of eq.(3), the first order term of dI/dV reads

$$\frac{dI}{dV} \propto \varrho(\mathbf{R}, E_F + eV) \quad (4)$$

which is implicitly used for rather primitive interpretation of STM/STS. However, this simplified picture is not enough for a quantitative interpretation of the experimental data and fails to explain some important phenomena, such as the negative differential conductance[13]. To see the tip effect, it is crucial to make a first-principles calculations of the surface and the tip system. For the theoretical simulation we use slab models for the sample surface and the cluster models for the tip. The Greens functions for the both systems are calculated by the standard density functional approach(LDA). The tunnel current is numerically evaluated by the expression (1),(2).

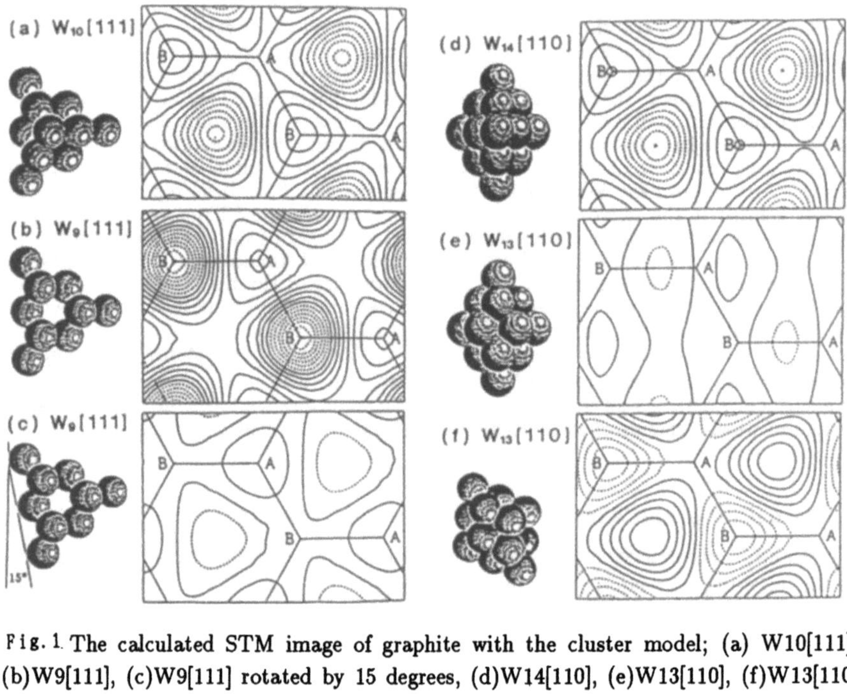

Fig. 1 The calculated STM image of graphite with the cluster model; (a) W10[111], (b)W9[111], (c)W9[111] rotated by 15 degrees, (d)W14[110], (e)W13[110], (f)W13[110] rotated by 30 degrees. The top views of the tip are shown in the left hand side

3. Simulation of STM/STS of graphite surface

By the tips W10[111], W14[110] with a single apex atom, STM images as shown in Fig.1a and 1d result.[6] In these calculations The tip-surface distance is assumed as 2.6A and the tip bias voltage is taken as -0.5V. The theoretical STM images agree well with experimental obsevation. The strong current region is located on the B site of graphite, and forms a triangular lattice as a whole. This is because the wavefunctions with the energy around the Fermi level have large amplitudes around the B site and not on the A site. In the image of Fig.1b, which is obtained by the model tip W9[111], the correspondence between the current distribution and the lattice structure is shifted from that for the normal image, Fig.1a,d. This can be explained as follows. In the tip geometry of Fig.1b, the three top W atoms come very close to one of the B site, when the tip axis is located on the A site. Thus the total tunnel current becomes maximum at this tip position.

Fig.1c shows the simulated STM image of graphite when the tip is rotated by 15 degrees from the geometry of Fig.1b.[6] The STM image changes significantly from that of Fig.1b. On the other hand for the tip with a single apex atom, the STM image hardly changes with the rotation of the tip. These features indicate that the tunnel current is really concentrated on the single apex atom. Fig.1e, and f are obtained by the tip W13[110], which has four W atoms on the top plane of the tip. The STM image of Fig.1e shows rather abnormal pattern flowing in the vertical direction. The origin of the abnormal image is an interference effect of tunnel current elements through the different atoms on the top plane of the tip. The abonormal image sensitively changes with the rotation of axis as shown in Fig.1f.

4. The STM Image of Si(111)$\sqrt{3} \times \sqrt{3}$-Ag Surface

The STM image for negative tip bias by van Loenen et al[14] revealed the distribution of the bright spots arranged in a honeycomb lattice pattern. However the naive interpretation of the STM image assigning the bright spots either to Si atoms or Ag atoms failed to reproduce results consistent with other experiments. On the other hand, the modified HCT model[15] based on CAICISS or X-ray standing wave method seemed promissing, but disagreement with the STM image has been mysterious.

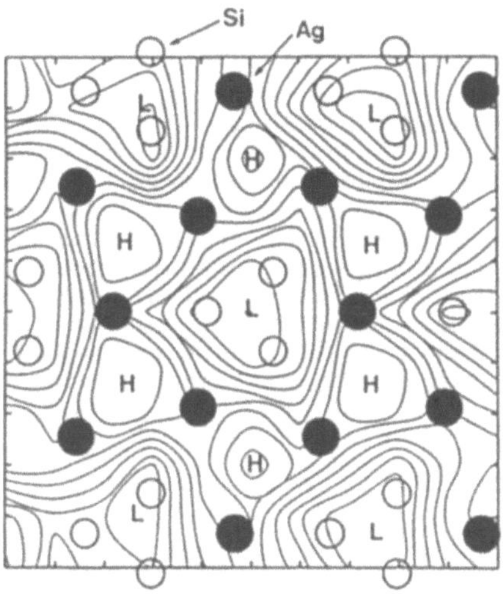

Fig. 2 The calculated current contour map in logarithmic scale for the Si(111) 3x 3-Ag HCT model. The tip W10[111] is located at 3.7A above the surface and the tip bias is -1.0V.

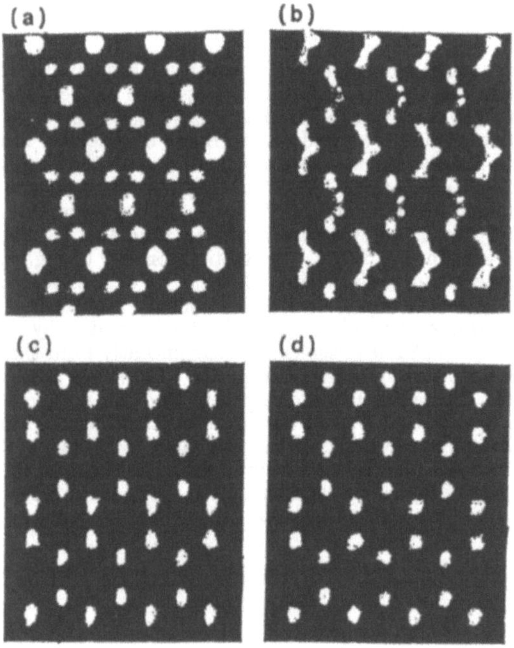

Fig. 3 The calculated STM image of the Si(111) 3 x 3 -Ag HCT surface by the tip W13[110] with the tilting angle (a) 0, (b) 10, (c) 20 and (d) 30 degrees. The tip height and the tip bias are the same as those in Fig.2

Watanabe et al[9] tried the theoretical simulation of the STM image of this surface, and obtained a surprising agreement between the simulated STM image and that actually observed. The reason why the HCT model gives the honeycomb like STM image is understood by Fig.2. The bright spots do not correspond to any of the atoms, but located at the inner region of the triangles formed by the surface trimer Ag atoms. The simulated STM image for the positive surface bias showed a reticulate structure, which reproduced well recent experimental results[10]. The theoretical simulations mentioned above were performed with the tip model W10[111]. Almost the same images have been obtained by W14[110] tip model. Thus we obtain the normal images by the tip models with a single apex atom. But if we use the tip model W9[111] or W13[110], which has more than one atom on the top plane, the obtained images are quite different from the normal one.[11] However interesting enough, these abnormal images quickly recover to the normal one, when the tip axis is tilted from the surface normal. One of the examples is shown in Fig.3. Tilting by about 20 degrees is enough to obtain the normal image.[16] This feature is explained by the concentration of the tunnel current to one of the four W atoms on the top tip plane.

5. Si(100) reconstructed surface and its step structure

The problem of this surface is that the symmetric dimers have been dominantly observed in the defect free region of the terrace[17], whereas the most stable structures theoretically predicted[18] is the asymmetric dimer structures. Other experimental evidences than STM also supported the asymmetric dimers for the low temperature phase[19]. The result of the theoretical simulation of the STM image assuming the symmetric dimers

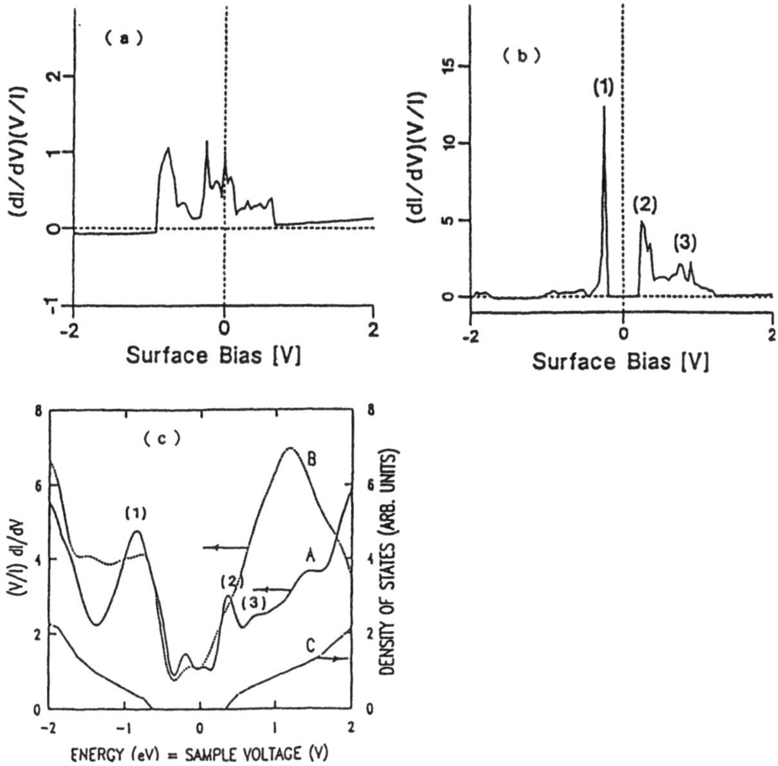

Fig. 4 Simulated STS spectra for (a) the symmetric dimer Si(100) 2x1 surface, (b) for the asymmetric dimer Si(100) c(4x2) surface compared with the experimental observation (R.J.Hamers, Ph. Avouris and F. Bozso, Phys. Rev. Lett. 59 (1987) 2071) (see Curve A)

at first sight reproduced well the experimental feature[5,20]. But the calculated STS spectrum for this surface showed the metallic character, as seen in Fig.4a, which is in contradiction with the experiment[21]. On the other hand as shown in Fig.4b, the simulated STS spectrum for the asymmetric dimer c(4x2) surface reproduced well the experimental results, which were observed at the seemingly symmetric dimer region. Therefore it has been speculated that the seemingly symmetric dimer is caused by the time averaged feature of the flip-flop motion of the asymmetric dimers. Very recently this speculation is actually confirmed by the STM observation at a lower temperature,T=120K by Wolkow[22]. Namely, the seemingly symmetric images gradually changed to the zigzag pattern due to the freezing of the flip-flop motion.

The steps of Si(100) surfaces have been clearly observed by STM[23], and its microscopic properties are very important for the study of surface dynamical processes. We investigated the electronic structure of S_A and S_B step with the superlattice model shown in Fig.5[24]. In the central terrace of Fig.5, two dimer rows are arranged in the c(4x2) structure, while on the side terrace the dimer rows are truncated perpendicularly to the steps. It is found that the pseudo-one-dimensional bands of dimers in the c(4x2) region slightly overlap with the Fermi level. This is in contrast with the band structure in the perfect c(4x2) surface showing the semiconducting nature.[18] Fig.6 shows the theoretical STM image with the tip model W10[111] as shown in the upper inset. The tip height is 2.6A from the top most atom on the central terrace, and the surface bias is -1.0V. The

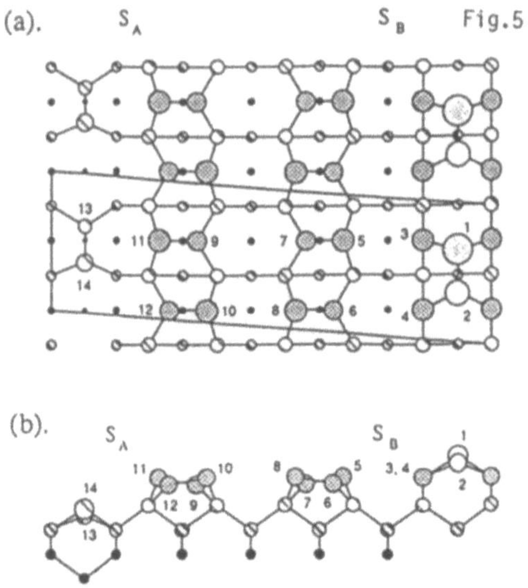

Fig. 5 The model of the step structure of Si(100) surface; (a) top view and (b) side view. The unit cell is marked by a parallelogram.

Fig. 6 The current contour map for the central terrace part of Fig.5. The S_A and S_B step appear at the upper and the lower region as marked by triangles.

up dimer atom on the terrace is images as a bright spot and this feature does not change by the reverse of the bias polarity. It should be noted that the second

layer atoms bonded to the dimer atom #11 and #12 at the side of the S_A step can not be distinctively seen, which is the same feature as that observed by the experiment. the reason is ascribed to the dull shape of the tip;the side atoms of the tip considerablly contributes to the tunnel current at the step.

6. Transparency of the STM image

Depending on the bias voltage the STM image of weakly adsorbed molecules becomes transparent[25]. This is a rather unusual behavior since usually the tunnel current is most sensitively influenced by the outermost structure. A similar strange behavior is the moire pattern formation between the outermost and the substrate surface[26]. The underling mechnism of these behaviors is not trivial. Recently we have proposed a theoretical model for these phenomena based on a transfer matrix formalism[10]. But we just confirm here the transparent STM images by the first-principles calculations of electronic states[10].

The first example is benzene molecules adsorbed on graphite substrate. We make the simulation with the model shown in Fig.7. In this model the molecules are arranged in $\sqrt{7} \times \sqrt{7}$ superstructure and placed at 3.35A above the graphite layer. Fig.8a and b show the STM images for the tip height 2.6A for the tip bias -1.0V and -0.1V, respectively. For the larger bias(Fig.8a), a doughnut like STM image is obtained. This corresponds to the electron cloud distributing along the benzene ring. On the other hand for the smaller bias voltage(Fig.8b), the alternate three carbon atoms are heighlighted. These carbon atoms corresponds to the B site of the graphite surface. The STM image of Fig.8b can be interpreted as the substrate graphite image which becomes visible through the benzene molecule. The calculated LDOS indicates that the heigest occupied molecular

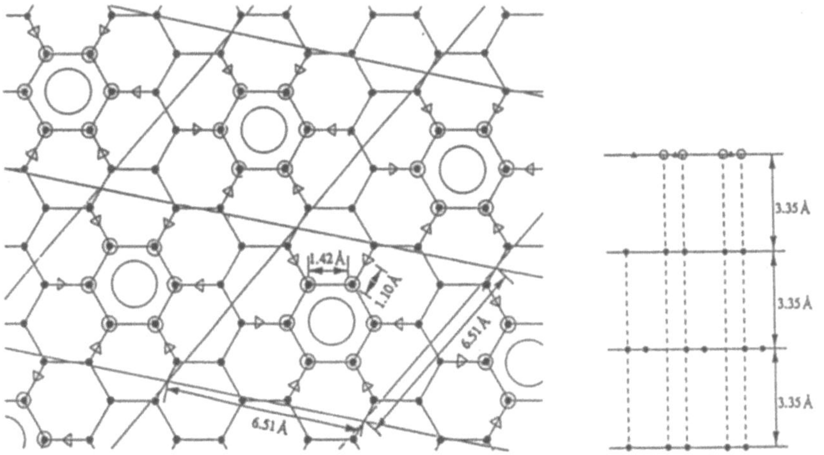

Fig. 7 The model of benzene molecules physisorbed on graphite surface: the left figure is the top view and the right is the side view. The open circle, triangle and filled small circle correspond respectively to C atom and H atom in benzene, and C atom of graphite.

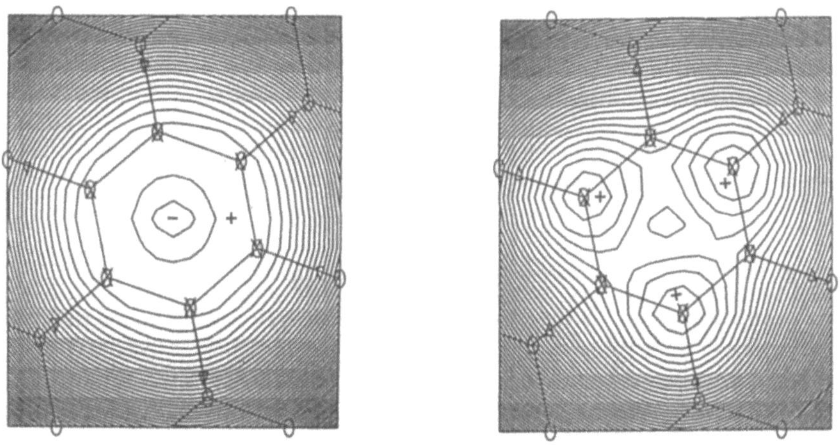

Fig. 8 Calculated current contour map for benzene molecule on graphite for the surface bias (a) 1.0V, and (b) 0.1V

level(HOMO) and the lowest unoccupied molecular level(LUMO) are apart from E_F by about 3eV. Therefore for the smaller bias the admixture of the benzene molecular orbital is very small and the tail of the wavefunctions are dominated by the substrate graphite chracter. On the other hand, if the energy of the tunnel electron is shifted towards HOMO or LUMO, the orbital admixture of benzene molecule becomes significant. The structure of the admolecule wavefunction becomes dominant in the vacuum tail, and this is the reason why the STM image represents the molecules for the higher bias voltages.

For the case of graphite monolayer on TiC(111) substrate, there observed two periods of the moire pattern[26]; the shorter one of the order of 5A, i.e., about twice as large as the graphite lattice vector, and the longer one with the order of 27A. The electronic state

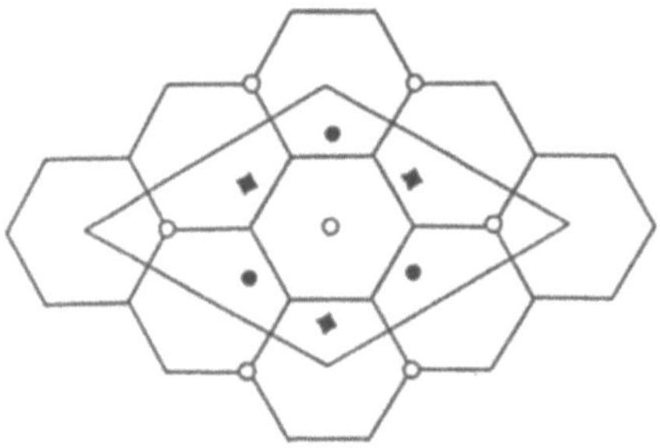

Fig. 9 The top view of the model of graphite monolayer formed on TiC(111) surface. The unit cell for the commensurate supercell is shown by a parallelogram.

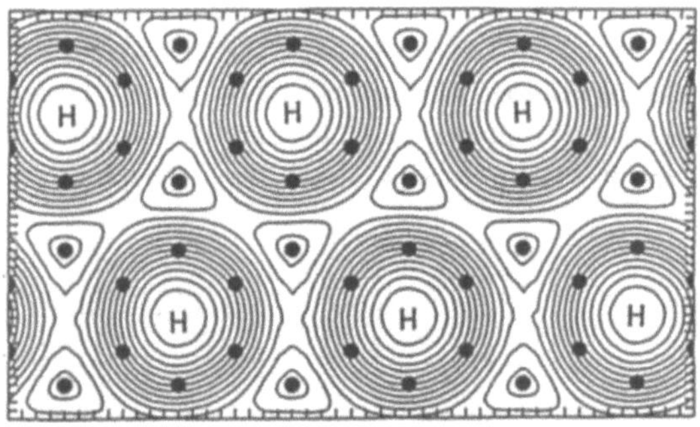

Fig. 10 The calculated corrent contour map for the graphite monolayer formed on TiC(111) shown in Fig.9. The filled small circle shows the C atom of graphite.

of the graphite/TiC(111) surface is calculated with the model as shown in Fig.9.[27] In this model, graphite monolayer is placed on the Ti surface of the TiC(111) slab which is formed by 6 atomic layers. In the actual system the graphite structure is incommensurate with that of the substrate. But here we focus on the dominant effect of the electronic interaction introduced by the substrate, so we elongated the graphite lattice constant slightly(by about 6%).

Fig.10 shows the calculated STM image of this surface[27] for the tip height 5.3Å from the top layer and the tip bias 0.5V. It is surprising that the STM image is quite differnt from that expected by the graphite layer struture. The strongest current region appears at the center of the carbon network hexagon sowing the 2x2 superlattice. None of the individual carbon atoms on the graphite layer can not been seen. The STM image reflects strongly the atomic structure of the substrate layer. The calculated STM image roughly reproduces the feature of the short period component of the experimental moire pattern.

7. Summary

A method of a theoretical simulation of STM images and STS spectra by the LDA method is presented. In this simulation, the real atomistic structure of the tip is taken into account and therefore the effect of the miscroscopic feature of the tip can be clarified. As typical examples, the STM images of graphite and Si(111) 3x 3-Ag surface, and Si(100) dimer structure and the step are calculated and compared with the experiments. It is found that the tip with only a single apex atom normal images with atomic resolution can be obtained. On the other hand, with the tip with more than one atom on the apex abnormal images resulted. This is due to the distribution of the tunnel currents over the apex atoms of the tip. Sometimes significant interference takes place among the tunnel current elements flowing through different atoms. If there is only one atom on the tip, the tunnel current tends to concentrate on this atom. This effect is confirmed by the fact that the abnormal image quickly recovers to the normal one, when the axis of the tip is tilted from the surface normal.

As an exotic phenomena observed by STM, we discussed the transparent image of weakly adsorbed molecules on a substrate by the example of graphite/benzene. The moire pattern formation between the graphite monolayer and the substrate is investigated in the same context. Depending on the bias voltages the substrate structure appears as the image. The reason of this is discussed elsewhere with analytical theory.[10]

Acknowledgement

The author would like to thank Mr.K.Kobayashi, Dr.T.Uchiyama, Dr.N.Isshiki, Dr.S.Watanabe, Dr.H.Kageshima for collaboration. The numerical calculation was performed using the HITAC S-820 computer system at the Computer Center of the University of Tokyo and the Computer Center of Institute for Molecular Science. This work was supported by a Grant-in-Aid from the Ministry of Education, Science, and Culture, Japan.

References

[1] "Scanning Tunneling Microscopy I,II,III" eds. H.J.Güntherodt and R.Wiesendanger (Springer, 1992)

[2] R.M.Feenstra, J.A.Stroscio and A.P.Fein, Surface Sci., 12 (1987)95

[3] M.Tsukada, K.Kobayashi, N.Isshiki and H.Kageshima, Surface Sci. Reports,33 (1991) 265

[4] N.Isshiki, K.Kobayashi and M.Tsukada, Surface Sci., 238 (1990) L439

[5] N.Isshiki, H.Kageshima, K.Kobayashi and M.Tsukada, Ultramicroscopy 42-44 (1992) 109

[6] T.Uchiyama and M.Tsukada, Surface Sci., in press

[7] S.Watanabe, M.Aono and M.Tsukada, Phys. Rev. B44 (1991) 8330

[8] S.Watanabe, M.Aono and M.Tsukada, Ultramicroscopy 42-44 (1992) 105

[9] S.Watanabe, M.Aono and M.Tsukada, Appl. Surface Sci., 60/61 (1992) 437

[10] M.Tsukada, N.Isshiki and K.Kobayashi, Surface Sci., in press.

[11] J.Bardeen, Phys. Rev. Lett., 15 (1961) 57

[12] M.Tsukada and N.Shima, J.Phys.Soc.Jpn., 56 (1987) 2875

[13] M.Tsukada and K.Kobayashi, Surface Sci., 242 (1991) 12

[14] E.J.van Loenen, J.E.Demuth, R.M.Tromp and R.J.Hamers, Phys. Rev. Lett., 58 (1987) 373

[15] M.Katayama, R.S.Williams, M.Kato, E.Nomura and M.Aono, Phys. Rev. Lett., 66 (1991) 2762

[16] S.Watanabe, M.Aono and M.Tsukada, Surface Sci., in press

[17] R.J.Hamers, R.M.Tromp, and J.E.Demuth, Phys. Rev. B34(1986)5343

[18] Z.Zhu,N.Shima and M.Tsukada, Phys. Rev. B40 (1989) 11868
[19] T.Tabata, T.Aruga and Y.Murata, Surface Sci., 179 (1986) L63
[20] H.Kageshima and M.Tsukada, Phys. Rev. B46 (1992)6928
[21] R.J.Hamers, Ph.Avouris and F.Bozso, Phys. Rev. Lett., 59 (1987) 2071
[22]R .A.Wolkow, Phys. Rev. Lett., 68 (1992) 2636
[23] F.Iwawaki, M.Tomitori and O.Nishikawa, J.Vac.Sci.Technol., B9(1991)711
[24] T.Uchiyama and M.Tsukada, Surface Sci., in press
[25] W.Mizutani, M.Shigeno, M.Ono and K.Kajimura, Appl. Phys. Lett., 56 (1990) 1974
[26] H.Itoh, T.Ichinose, C.Oshima, T.Ichinokawa and T.Aizawa, Surface Sci., 254 (1991) L437
[27] K.Kobayashi, Y.Souzu, N.Isshiki and M.Tsukada, Appl. Surface Sci. 60/61 (1992) 443.

Part III

Phase Stability and Kinetics in Alloys and Compounds

Order and Phase Stability in Alloys

F. Ducastelle

Office National d'Etudes et de Recherches Aérospatiales (ONERA), B.P. 72, F-92322 Châtillon Cedex, France

Abstract. A general theory of ordering effects on fixed lattices is presented. Effective multiatom interactions are derived from the electronic structure and are shown to account for many observed ordering trends.

1. Introduction

Due to the presence of interatomic interactions, any alloy should eventually order or phase separate at low temperature. This is observed in many systems. When the ordering interactions are strong, the alloy may remain ordered up to the melting point (Ni_3Al for instance). When they are very weak, the solid solution can be the only observable phase since at low temperature atomic diffusion is no longer efficient (Cu-Ni). In the intermediate regime we have a more interesting situation with one or several ordered phases at low temperature which disorder at a critical temperature before melting (Cu-Au, Pd-V, Ni-Fe, etc.). In the simplest case all phases are built on a fixed underlying lattice such as the FCC or BCC lattice, but more frequently, several structures are involved and many compounds display structures which are not observed in elemental metals (Laves phase, σ phase, A15 phase, etc.).

It is of great interest, both from a fundamental and a practical point of view to understand what stabilizes a particular structure at a given concentration and temperature and then to predict the phase diagrams of alloys.

Up to recent years, most theories in this field were based on phenomenological models. The stability of several crystalline structures has been interpreted using geometrical arguments by looking for the most efficient way of filling space with hard spheres of different radii. Size effects are also known to account for the limits of stability of solid solutions. Chemical or electronic parameters have also been used. One of the celebrated Hume-Rothery rules for instance relates the stability of several structures to definite values of the electronic ratio (number of valence electrons per atom). Finally various electronegativity scales have been introduced. As far as phase diagrams are concerned, most approaches have been based on phenomenological thermodynamic models using regular solution models or various improvements on them.

Recent works indicate that it is now possible to tackle these problems from microscopic theories based on first principles. This confidence is based on the success of first principles electronic structure calculations concerning elemental

Springer Series in Solid-State Sciences, Volume 114
Interatomic Potential and Structural Stability Eds.: K. Terakura and H. Akai

metals as well as intermetallic compounds and on the development of efficient thermodynamic tools. In fact "ideal" methods starting from the Schrödinger equation and from general thermodynamics are generally not so useful when one is interested in a physical understanding of fairly complex systems, and it is still quite valuable to define intermediate models containing a few relevant parameters.

Many modern studies use an approach similar to that presented in this short review which can be summarized as follows. We start from electronic structure models based on various well defined approximations: the one-electron approximation based on the so-called local density approximation, the tight-binding approximation when necessary, etc. These approximations are now well controlled in the case of elemental metals or simple ordered compounds and provide total energies with a very good accuracy. The next step is to calculate the energy of any atomic configuration which obviously cannot be done exactly. At this stage it is shown that this energy can be obtained through a generalized perturbation expansion starting from the disordered state.

At the end of this process, the electronic degrees of freedom are eliminated and we are left with the desired expression for the energy of any configuration as a function of pair and multiplet interactions. In the case of a binary alloy we recover the celebrated Ising model or direct generalizations of it, were it not for the concentration dependence of the interactions which is completely unavoidable in the theory. The implications of this dependence can be considerable, in particular when studying phase diagrams (principally because of the common tangent rule), but they are still far from being well characterized, both experimentally and theoretically, and it is hoped that at least locally on the concentration axis the previous theory offers a good justification for the use of Ising models.

The next step is therefore a thermodynamic one and consists in exploring the properties of the Ising model, i.e. to study the nature and the stability of ordered structures as a function of concentration, temperature and of the interactions. Exact results can be obtained at zero temperature but approximations are required at finite temperature. In the case of short-range interactions the modern developments of generalized mean field methods such as the cluster variation method now provides very accurate results, as tested by Monte Carlo simulations.

We are thus able in principle to go from electronic structure calculations to phase diagrams through a set of well controlled approximations. At least this is so when ordering effects can be assumed to take place on a fixed, rigid lattice. These restrictions are rather important, since they are necessary in order to speak of Ising models with discrete numbers of degrees of freedom per site. However even if this has still not been worked out in detail there is no difficulty in principle in extending the whole chain of arguments to include elastic and relaxation effects; it is also be possible to compare ordered structures on different types of lattices, i.e. to treat structural effects.

The present paper is somewhat schematic. Details and relevant references can be found in a recent book [1].

2. Tight-binding model for alloys

The tight-binding formalism is particularly well adapted to a semi-quantitative discussion of strong alloying effects. It also applies quite naturally to transition alloys. Furthermore there is no difficulty in principle, if not in practice, to derive the appropriate extensions involving full band structure calculations.

We consider here binary alloys $A_{1-c}B_c$ on fixed underlying lattices. For convenience we assume crystalline structures with a single atom per unit cell (e.g. fcc or bcc). Each atomic configuration is characterized by the value of the so-called occupation numbers p_n^i equal to unity or zero depending on the presence or not of an atom of type i at site n ; obviously $p_n^B = 1 - p_n^A$. The simplest tight-binding hamiltonian H can then be written

$$H = \sum_n |n> \varepsilon_n < n| + \sum_{n \neq m} |n> \beta_{nm} < m|,$$

where $|n>$ is an atomic state centred on site n and ε_n , β_{nm} are the corresponding atomic level and hopping integrals, respectively,

$$\varepsilon_n = \sum_{i=A,B} p_n^i \varepsilon^i ,$$

$$\beta_{nm} = \sum_{i,j=A,B} p_n^i p_m^j \beta_{nm}^{ij} .$$

In the case of transition alloys the degeneracy of the d states has to be included, but for simplicity we shall drop here the corresponding indices which can easily been reintroduced when necessary. In this simple scheme, the electronic structure of the alloy is then characterized by a few parameters. Taking as the origin of energy the mean atomic level $\bar{\varepsilon} = (1-c)\,\varepsilon^A + c\,\varepsilon^B$ a first one is the so-called diagonal disorder parameter $\delta = \varepsilon^B - \varepsilon^A$ which plays the part of an electronegativity difference. The off-diagonal disorder is then related to the differences between the hopping integrals. A very convenient and reasonable approximation is to set $(\beta^{AB})^2 \approx \beta^{AA} . \beta^{BB}$. Since the widths W^A and W^B of the pure elements are proportional to β^{AA} and β^{BB} respectively, off-diagonal disorder can be characterized by the ratio $(W^A - W^B)/\overline{W}$ where $\overline{W}$ is the mean band width. The bandwidth of transition metals decreases along a transition series and increases from the 3d to the 5d series with extreme values of the order of 4 and 8eV. The energy levels on the other hand decrease about one eV per element (see Fig.1).

In many cases then, off-diagonal disorder can be neglected (a typical exception would be Zr-Ni alloys) and the principal parameter is the dimensionless ratio $\delta/\overline{W}$ which is at most of the order of unity for alloy constituents with very different numbers of d electrons N_e^A and N_e^B . When $\delta/\overline{W} << 1$ standard perturbation theory applies, and to lowest order the density of states is that of an effective metal of bandwidth $\overline{W}$ and centred on the mean level $\bar{\varepsilon}$. In the opposite limit we have a split-band regime with subbands containing states extending principally over A or B atoms. From very general arguments it is then easy to determine the qualitative

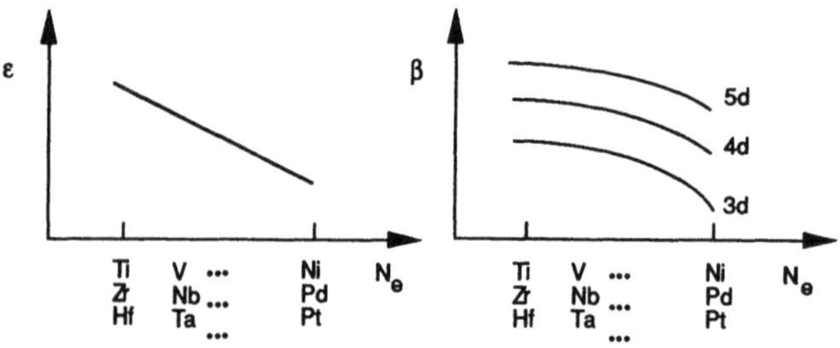

Figure 1. Typical variations of the atomic levels and of the hopping integrals as a function of the number of electrons.

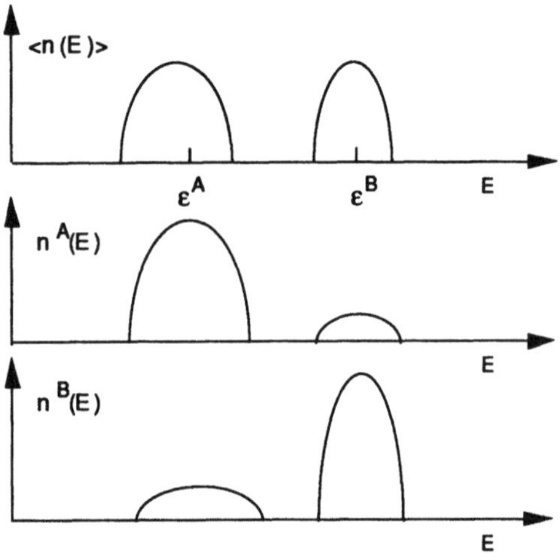

Figure 2. Total and partial densities of states.

shape of the density of states as a function of the degree of order and of the concentration [1,2]. We can define an average density of states <n(E)>, averaged over all configurations of concentration c and local conditional densities of states $n^i(E)$, i = A,B averaged over all configurations where the occupation of a given site is held fixed. Typical variations of these densities of states are sketched in Fig.2.

Local charges are obtained by integrating the conditional densities of states up to the Fermi level. In general we therefore obtain charge transfers. At this point some self-consistency rule is necessary since for example the diagonal disorder parameter δ has been considered up to now as an atomic quantity. The simplest rule is to postulate local neutrality. This implies that δ has to be self-consistently recalculated. In general, this yields values smaller than those deduced from the free atom energy levels; this is a typical feed-back effect. Notice that global neutrality is always insured, by definition, since it is used to locate the Fermi energy.

3. Effective multiatom interactions

When local neutrality is preserved the total energy of an alloy is directly related to its band energy

$$U = \int^{E_F} dE\, E\, n(E) \quad ,$$

where E_F is the Fermi energy. A simple moment analysis yields very interesting informations on the behaviour of this quantity (or more precisely of the related grand-potential $\Omega = U - E_F \overline{N}_e$,) as a function of the filling of the band $\overline{N}_e$.

3.1 Moments of the density of states and effective interactions

The normalized n^{th} moment of the density of states μ_n is given by

$$\mu_n = \int dE\, E^n n(E) = \frac{1}{N_a} \mathrm{Tr}\, H^n \ ,$$

where N_a is the number of atoms. Using the decomposition $H = H^0 + V$ where H^0 is the off-diagonal part of H involving transfer integrals and V its diagonal part, it is straightforward to realize that the first moment that depends on the atomic configuration at fixed concentration is the fourth one, the part depending explicitly on this configuration $\Delta\mu_4$ being proportional to $\mathrm{Tr}\ H^0 V H^0 V$. Associating as usual with μ_n closed circuits of n elementary steps, we have here two types of steps: those involving the diagonal matrix elements represented by closed loops and those involving the hopping integrals associated with interatomic jumps. This yields

$$\Delta\mu_4 = \frac{2}{N_a} \sum_{n,m} (p_n - c)(p_m - c)\, \beta_{nm}^2\, \delta^2 .$$

It can be shown on the other hand that the moments Δm_n of the ordering energy considered as a function of the Fermi energy are proportional to $-\Delta\mu_{n+2}$. Thus $\Delta m_0 = \Delta m_1 = 0$, Δm_2 is negative and related to pair interactions. More precisely one can defined moments of pair and cluster interactions and the above equation tells us that the second moment of the pair interaction V_{nm} is proportional to $-\beta_{nm}^2 \delta^2$. In general we only keep hopping integrals connecting an atom and those of its first coordination shell (first neighbours in the fcc lattice, first and second neighbours in the bcc lattice) so that only first neighbour (in this sense) pair interactions V_1 contribute to Δm_2. The values of the first moments of $V_1(E_F)$ provide us with very useful sum rules from which we deduce that $V_1(\overline{N}_e)$ has in general two zeros and varies as indicated in the upper curve of Fig.3.

This shows that ordering (phase separation) is favoured for intermediate (extreme) fillings of the d band. Considering higher order moments it can be realized that the most important contribution to the third moment of V_1 is equal to $-(1\text{-}2c)\, \beta_{nm}^2 \delta^3$ which induces the asymmetry sketched in this figure.

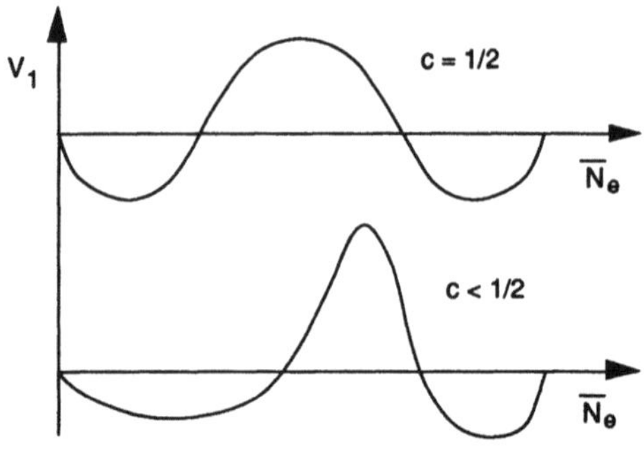

Figure 3. Schematic variations of V_1 as a function of $\overline{N}_e$.

Considering now further pair interactions, it is clear that they involve higher order moments. For example, in the fcc lattice we have a second class of neighbours, from the second to the fourth one, which can be reached in two steps. Their first non-vanishing moment , related to $\Delta\mu_6$ is the fourth one. We therefore expect $V_{2,3,4}$ to be of the same order of magnitude. To be more precise one has to calculate the number of circuits involving a given pair, and also to take into account the fact that straight self-retracing paths contribute more than open ones. These effects have a tendency to cancel each other so that we predict the hierarchy $|V_1| >> |V_2, V_3, V_4| >> |V_5,...|$. Similar arguments allow us to guess the hierarchy involving higher order cluster interactions. Once again the most important contributions are associated with straight paths. For example the most important triplet in a fcc lattice is that involving the first and fourth neighbours and may not always be negligible, compared to the pair interactions $V_{2,3,4}$.

3.2 Generalized perturbation method and discussion

All these arguments can be made more quantitative by using a more detailed theory where the band energy of a particular atomic configuration is compared to that of the fully disordered state and then expanded in terms of multiatom interactions (generalized perturbation method or GPM). In practice the disordered state is treated within the coherent potential approximation (CPA). Details can be found elsewhere [1,2]. The results for the pair and cluster interactions are in complete agreement with the predictions of the moment analysis (see Fig.4) and show that in most cases the generalized perturbation expansion is rapidly convergent. Note that these interactions depend on the concentration through the variations of the Fermi energy and of the CPA disordered medium.

Several effects neglected in this discussion (dependence on the crystalline structure, off-diagonal disorder, self-consistency, magnetism, etc.) can be treated within the same scheme.

The method presented previously is to be contrasted with another frequently used method which may be called an inverse method. The idea is that if the ordering energy is considered as a (linear) function of a finite set of interactions, then these interactions can be determined from the calculation of the ordering energies of a

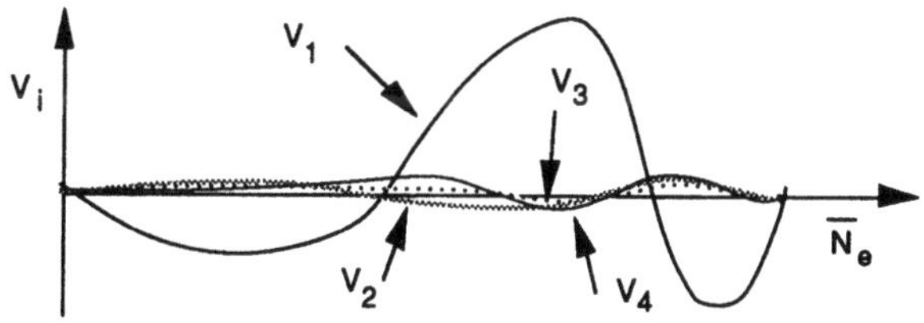

Figure 4. Pair interactions on the fcc lattice (c=1/4, $\delta/\overline{W}$ = 0.45); after Bieber [9].

sufficiently large number of ordered structures. Since this can be achieved with a very high accuracy, using first-principles band structure calculations, it is sometimes argued that this is *the* way to proceed. There are however clear disadvantages with this approach; first the relevant interactions are at best determined using trial and error methods, but above all the physical interpretation of the variations of the interactions with various electronic parameters can be completely lost.

Inverse and direct methods have been compared in detail by Sluiter and Turchi [3]; see also Ferreira et al. [4], Wei et al. [5]. A comparison with different but similar perturbation expansions is also given in Gonis et al. [6]. It is also possible to replace the CPA by numerical averages over configurations [7]. The concentration dependence of the interactions has also been discussed in this framework [8]. Many results of these theories are very similar, which is quite easy to understand using the general arguments presented above.

4. Effective Ising model and phase stability

For a given atomic configuration characterized by the occupation numbers $p_n = p_n^B$ the energy $U(\{p_n\})$ can therefore be written

$$U(\{p_n\}) = \frac{1}{2} \sum_{n,m} V_{nm}\, p_n p_m + \cdots ,$$

where constants depending only on the (fixed) concentration have been omitted. This is an effective Ising model with concentration dependent interactions. If this dependence is neglected, standard methods for describing the thermodynamics of this model can then be used. We shall just discuss here a few applications to transition alloys, principally at zero temperature.

Transition alloys are well described qualitatively within the tight-binding scheme. We recall that the important alloy parameters are the alloy parameter $\delta/\overline{W}$ and the filling of the d band $\overline{N}_e$. Since δ is itself approximatively proportional to $\Delta N_e = N_e^B - N_e^A$, it is quite natural to classify the structures by using the coordinates ΔN_e and $\overline{N}_e$ [10]. Such maps are very successful indeed in that the different ordered structures generally belong to different non-overlapping domains.

We comment here on the the map at $c = 1/4$ [1,10,11] and consider the simplest structures, i.e. the familiar $L1_2$ and DO_{22} structures built on the fcc lattice as well as the equivalent structures on the hcp (DO_{19} and DO_a) or double hcp lattice (DO_{24}), and the A15 structure.

The observed ordering tendencies are quite clear. A15 appears when $\Delta N_e > 0$ and all other non-magnetic structures when $\Delta N_e < 0$. Note that our sign convention is such that for instance ΔN_e is positive for PtV_3 but negative for Pt_3V (we always assume a compound A_3B). It is also striking that all DO_{22} structures, plus the "equivalent" DO_a structure occur at the same point, $\overline{N}_e = 7.75$, $\Delta N_e = -5$ and therefore correspond to elements of the column of vanadium and of that of nickel. The non-magnetic $L1_2$ structures are all in the range $6.5 \leq \overline{N}_e \leq 7.5$ whereas most equivalent hexagonal structures occur for $\overline{N}_e = 7.25$.

From the theoretical side, consider first the fcc lattice. We know that in general $|V_1| >> |V_2, V_3, V_4| >> | V_5, V_6, ...|$, so that ordered structures only appear when $V_1 > 0$. Calculating V_1 from the electronic structure defines a domain in which non-magnetic systems may order. Now, on the fcc lattice and for $c = 1/4$, the only possible ground states in the presence of interactions up to fourth neighbours are $L1_2$ and DO_{22} . The two structures have identical numbers of AB first neighbour pairs, which means that the difference in energy between them is much smaller than V_1 . It is given by the combination $\xi = V_2 - 4V_3 + 4V_4$, and moment arguments show that ξ should have at least four zeros as a function of band filling. Tight-binding calculations then clearly show a stability exchange between $L1_2$ and DO_{22} for $\overline{N}_e$ about seven, in good agreement with the observed structures [12]. Large values of $\overline{N}_e$ are irrelevant (since there is no order) as well as small ones (since the lattice is no longer FCC). These results have been confirmed by more sophisticated electronic structure calculations[13].

There are numerous other examples showing that simple theoretical arguments predict the correct structural trends. For example the hierarchy between the interactions on a fcc lattice shows that it is generally necessary to include third and fourth neighbour interactions as soon as second neighbour interactions are introduced. This is frequently verified, in the case of transition alloys at least. Actually many observed ordered structures require at least fourth neighbour interactions to be stabilized. Conversely a structure like that of CuPt ($L1_1$) requires second neighbour interactions comparable with the first neighbour ones and is in fact never observed except precisely in the case of CuPt!

As another example illustrating how we have elucidated trends, consider the nickel-based transition alloys, in particular their phase diagrams as compiled by De Boer et al.[14]. When ΔN_e is small, ordering is absent except in Pt-Ni . Small values of ΔN_e imply small, almost unobservable transition temperatures, and anyway large values of $\overline{N}_e$ induce a tendency to phase separation. For the case of 4d and 5d alloying elements, except for Pt-Ni, there is indeed no ordering tendency up to Tc and Re. Ordering effects in Ni-Fe and Ni-Mn are clearly correlated with the appearance of magnetism. Weak ordering effects appear when ΔN_e increases. Ni-Mo and Ni-W present ordered phases whereas SRO effects have clearly been detected in Ni-Cr . This is consistent with the fact that the order of magnitude of ordering energies should increase with the bandwidth, i.e. when going from the 3d to the 5d series. The same phenomenon exists for the vanadium column: on the Ni-rich side of the diagrams there are ordered structures on the fcc lattice which disorder at high temperature in the case of Ni-V but not in the case of Ni-Nb and Ni-Ta . Finally many very stable ordered compounds appear when ΔN_e is equal to 6 and 7 .Thus, except for the ordering in Pt-Ni which is still a subject of controversy, almost all is understood qualitatively for the case of Ni-based alloys. This is certainly a somewhat favourable situation, but many other successful analyses have been made. For example, the fact that off-diagonal disorder induces phase separation is well verified in the case of Cr-Mo and Cr-W alloys.

It is more difficult to calculate phase diagrams than to characterize the single phases, one reason being that it is not sufficient to know the interatomic chemical interactions. The free energy of the disordered state is also required. Although it can be calculated in principle, using the CPA for example, additional contributions related to elastic, vibrational and other effects have to be included. Another serious difficulty is that real phase diagrams frequently involve different crystalline structures, and at the moment there is no elaborate theory to describe the interplay between structural and chemical (ordering) effects. In practice, ordering effects are treated on fixed lattices and the free energies for different structures are compared. This is a correct procedure but the uncertainties concerning the structural energies or entropies are difficult to estimate, and generally some parameters have to be fitted to experimental data in order to produce reasonable phase diagrams. It is clear also that it is difficult to treat the liquid state on the same footing than the solid phases.

Despite all these problems several phase diagrams have now been calculated. There is no room here to undertake a comprehensive review of these results. We just mention some calculations concerning the Ti-Rh [15] Ni-Cr [16] Pd-Rh, Al-Li [17] phase diagrams and the recent detailed discussions concerning the Ni-Ti, Ni-Al and Ti-Al phase diagrams [18,19, 20].

References

[1] F.Ducastelle, *Order and Phase Stability in Alloys,* (North-Holland, Amsterdam, 1991).

[2] F.Ducastelle, in *Alloy Phase Stability,* edited by G.M.Stocks and A.Gonis, NATO-ASI Series E, Vol.163 (Kluwer, Dordrecht, 1989), p.293, and in *Electron Theory in Alloy Design,* edited by D.G.Pettifor and A.H.Cottrell (The Institute of Metals, London 1992), p.122.

[3] M.Sluiter and P.E.A Turchi, Phys. Rev. **B40**, 11 215 (1989).

[4] L.G.Ferreira, S.H.Wei and A.Zunger, Phys.Rev. **B40**, 3197 (1989).

[5] S.H.Wei, L.G.Ferreira and A.Zunger, Phys. Rev. **B41**, 8240 (1990).

[6] A.Gonis, X.G.Zhang, A.J.Freeman, P.Turchi, G.M.Stocks and D.M.Nicholson, Phys.Rev. **B36**, 4630 (1987).

[7] H.Dreysse, A.Berera, L.T.Wille and D.De Fontaine, Phys.Rev. **B39**, 2242 (1989).

[8] M.Asta, C.Wolverton, D.De Fontaine and H.Dreysse, Phys.Rev. **B44**, 4907 (1991).
C.Wolverton, M.Asta, H.Dreysse and D.De Fontaine, Phys.Rev. **B44**, 4914 (1991).

[9] A.Bieber, Thèse de Doctorat d'Etat, Université Louis Pasteur, Strasbourg, (1987).
[10] A.Bieber and F.Gautier, Acta Metall. **34**, 2291 (1986).
[11] M.Sluiter, P.Turchi and D.de Fontaine, J. Phys. F: Metal Phys. **17**, 2163 (1986).
[12] A.Bieber, F.Ducastelle, F.Gautier, G.Tréglia and P.Turchi, Solid State Commun. **45**, 585 (1983).
[13] S.Pei, T.B.Massalski, W.M.Temmerman, P.A.Sterne and G.M.Stocks, Phys.Rev. **B39**, 5767 (1989);
W.Lin, J. Xu and A.J.Freeman, Phys.Rev. **B45**, 10 863 (1992);
C.Wolverton, G.Ceder, D.de Fontaine and H.Dreyssé, Phys.Rev. **B45**, 13 105 (1992);
A.Pasturel, private communication (1992).
[14] F.R.De Boer, R.Boom, W.C.M.Matens, A.R.Miedema and A.K.Niessen, *Cohesion in metals: Transition Metals Alloys*, (North-Holland, Amsterdam, 1991).
[15] M.Sluiter, P.Turchi, F.Zezhong and D.de Fontaine, Phys. Rev. Letters **60**, 716 (1988).
[16] N.C.Tso, M.Kosugi and J.M.Sanchez, Acta Metall. **37**, 121 (1989).
[17] D.D.Johnson, P.E.A.Turchi, M.Sluiter, D.M.Nicholson, F.J.Pinski and G.M.Stocks, Mat. Res. Soc. Symp. **186**, 21 (1991);
A.Gonis, P.E.A.Turchi, M.Sluiter, F.J.Pinski and D.D.Johnson, Mat. Res. Soc. Symp. **186**, 89 (1991).
[18] D.H.Le, C.Colinet, P.Hicter and A.Pasturel, J. Phys.: Condens. Matter **3**, 7895 and 9965 (1991).
[19] A.Pasturel, C.Colinet, A.T.Paxton and M.van Schilfgaarde, J. Phys.: Condens. Matter **4**, 945 , (1992);
C.Colinet, P.Hicter and A.Pasturel: Phys. Rev. **B45**, 1571 (1992).
[20] M.Asta, D.de Fontaine, M.van Schilfgaarde, M.Sluiter and M.Methfessel, Phys.Rev. **B46**, 5055 (1992).

From Electronic Structure to Phase Diagrams

J.M. Sanchez and J.D. Becker

Center for Materials Science and Engineering,
The University of Texas, Austin, TX 78712, USA

Abstract. A first-principles statistical mechanics theory of alloy phase stability is presented. The theory incorporates *ab-initio* electronic total energies, configurational entropies and vibrational modes into the total free energy of disordered systems and intermetallic compounds. Applications of the theory are given for all three binary alloys in the Ru-Zr-Nb system using the Linear Muffin-Tin Orbital method for the total energy calculations, the Cluster Variation method for the description of the configurational entropy, and the Debye-Grüneisen approximation for the vibrational modes. The results are used to compute the equilibrium phase diagram for the Zr-Nb system and to assess current experimental data on phase stability in the Ru-Nb system. In the latter case, the calculations strongly suggest the DO_{19} structure as a likely candidate structure for the experimentally observed hexagonal-based compound Ru_3Nb, and the $L1_0$ structure as a likely candidate for the observed tetragonal phase near 1:1 stoichiometry.

1. Introduction

At present, a number of physical properties such as cohesive energies, elastic moduli and expansion coefficients of elemental solids and intermetallic compounds can be routinely calculated from first principles using, as input, only the atomic numbers of the constituent elements and the crystal structure of the solid. These achievements are a direct consequence of a mature theoretical and computational framework in solid state physics which has been in place for some time. Here we will focus on recent theoretical and computational developments in alloy theory which make possible the non-empirical calculation of phase diagrams and, in general, the study of phase stability strictly from first principles. Early studies of alloy phase equilibrium have shown that simple models are capable of reproducing quite well the most important features of alloy phase diagrams [1-5]. The essence of these models is the description of the energy in terms of pair and many-body interactions. In these developments, the Cluster Variation Method (CVM) of Statistical Mechanics [6] played a key role since it provided an efficient and computationally economical way of describing the configurational thermodynamic of alloys. The early success of the CVM calculations revived the old dream of computing alloy phase diagrams from first principles; i.e. from the knowledge of the electronic structure of the alloy.

One of the most significant developments in alloy theory, density functional theory and its computational version, the local-density approximation (LDA) [7], was fully developed and ready to be applied together with the statistical models to tackle the delicate problem of alloy stability at finite temperatures. This local-density approximation has been used repeatedly to calculate the total energy at zero

Springer Series in Solid-State Sciences, Volume 114
Interatomic Potential and Structural Stability Eds.: K. Terakura and H. Akai

temperature of pure solids [8-10], relatively simple compounds [11,12] and disordered alloys [13-15]. In general, results of the calculations reproduce physical properties within a few percent of the experimental values.

Other developments in the statistical thermodynamics of alloys led to the description of the energy in terms of multisite correlation functions [1,16]. In this cluster expansion, the energy is given by pair and many-body interactions. An attractive feature of this cluster representation for the configurational energy is that the alloy problem becomes essentially isomorphic to a generalized Ising model where the chemical interactions are relatively short ranged. Although these interactions include, in general, many-body terms as well as temperature and volume dependence, they can be easily treated using the CVM to calculate configurational free energies and, from them, the solid state portion of phase diagrams. The success of this cluster representation for the energy of alloys led Connolly and Williams [17] to propose the use of *ab-initio* total energy calculations of ordered compounds in order to obtain the set of effective pair and multisite chemical interactions. The proposal of Connolly and Williams clearly reflected the degree of confidence with which total energies could be calculated using the local-density approximation, and opened the door to first principles calculation of phase diagrams.

Among the first applications of this first-principles approach were studies of temperature-composition binary phase diagrams of noble-metal alloys [18] and semiconductor alloys [19]. Subsequently, numerous other cases have been investigated with relatively good results [20-27]. In general, the cluster expansion of the configurational energy converges relatively fast, thus offering a practical method for the determination of the chemical interactions in a disordered alloy system from total energy band calculations for a relatively small set of ordered compounds.

In the next section we present a cluster theory of the configurational thermodynamic of alloys. The description provides the formal framework for the treatment of short-range order (SRO) effects in the configurational energy and in the configurational entropy. The contribution to the free energy due to vibrational modes is also discussed. We make contact with microscopic electronic theories via the Linear Muffin-Tin Orbital (LMTO) approximation, which is used to calculate the total energies of selected compounds in the Ru-Zr-Nb system. The first principles statistical thermodynamic theory, which includes configurational, vibrational and volume relaxation effects, is applied to the computation of the equilibrium solid state phase diagrams of the Zr-Nb system. The electronic structure calculations are also used to shed light on conflicting experimental reports on the phase stability of the Ru-Nb system.

2. Configurational Thermodynamics

In this section we review briefly the general formalism used in the description of the configurational thermodynamics of alloys. The main result is the cluster expansion which provides the basis for the description of disordered alloys from the knowledge of the energy (binding curves) of ordered compounds. For the sake of simplicity we consider only binary systems although the theory can be easily extended to multicomponent alloys. [16]

2.1 Cluster Expansion

The configuration of a crystalline binary alloy is described in terms of spin or occupation numbers σ_i at each lattice site i which take, respectively, values +1 and -1 for components A and B. Any configuration of the system is then fully specified by the N-dimensional vector $\sigma=\{\sigma_1,\sigma_2...,\sigma_N\}$, where N is the number of lattice points. In general, one is confronted with the problem of describing functions that depend explicitly on the occupation variables σ_i, such as the energy of formation of the alloy. In order to provide an unambiguous description of such functions, it is convenient to introduce an orthogonal functional basis in configurational space. Although in the thermodynamic limit the dimension of the complete orthogonal basis is infinite, judicious choice of the basis functions allow us to obtain accurate approximations to the thermodynamic potentials in terms of subset of finite dimension.

For a single site i, the set of two polynomials in the discrete variable σ_i, namely the polynomial of order 0, $\phi_0(\sigma_i) = 1$, and the polynomial of order 1, $\phi_1(\sigma_i) = \sigma_i$, form a complete and orthonormal set, with the inner product between two functions of configuration $f(\sigma_i)$ and $g(\sigma_i)$ in the one-dimensional discrete space spanned by σ_i defined as:

$$< f(\sigma_i)\cdot g(\sigma_i) > = \frac{1}{2} \sum_{\sigma_i=\pm 1} f(\sigma_i)\, g(\sigma_i)\,. \tag{1}$$

The set of orthonormal characteristic functions in the N-dimensional discrete space spanned by the vector σ is obtained from the direct product of the $\{\phi_0(\sigma_i), \phi_1(\sigma_i)\}$, where i spans all crystal sites (i=1,2,...N). For a binary system, the resulting characteristic functions, $\Phi_\alpha(\sigma)$, are given by products of the spin operator σ_i over the sites of all possible clusters $\alpha=\{i_1,i_2,...\, i_n\}$ in the crystal [16]:

$$\Phi_\alpha(\sigma) = \prod_{i\in\alpha} \sigma_i = \sigma_{i_1}\, \sigma_{i_2} ...\sigma_{i_n}\,. \tag{2}$$

Accordingly, there is a one to one correspondence between the set of orthogonal functions $\Phi_\alpha(\sigma)$ and the set of all clusters α in the crystal, including the empty cluster for which $\Phi_0(\sigma) = 1$. The orthogonality of the characteristic functions $\Phi_\alpha(\sigma)$ is expressed by [16]:

$$\frac{1}{2^N} \sum_{\sigma} \Phi_\alpha(\sigma)\, \Phi_\beta(\sigma) = \delta_{\alpha,\beta}\,. \tag{3}$$

In view of Eqn. (3), any function of configuration, $F(\sigma)$, may be written as,

$$F(\sigma) = \sum_{\alpha} F_\alpha\, \Phi_\alpha(\sigma)\,, \tag{4}$$

where the sum extends over all clusters in the crystal, including the empty cluster, and where F_α is given by,

$$F_\alpha = < F(\sigma)\cdot\Phi_\alpha(\sigma) > = \frac{1}{2^N} \sum_\sigma F(\sigma)\, \Phi_\alpha(\sigma)\,. \tag{5}$$

Thus, the terms F_α are the projections of $F(\sigma)$ on the orthogonal cluster basis.

It should be noted that the space group symmetry of the crystal requires that the cluster projections F_α of the function $F(\sigma)$ be the same for all clusters α which are related by a symmetry operation (translation or point group). Accordingly, the cluster expansion in Eqn. (4) becomes:

$$F(\sigma) = \sum_{n=0}^{N} F_n\, \Theta_n(\sigma)\,, \tag{6}$$

where n labels the set of inequivalent clusters in the crystal. In the case of a disordered lattice, these cluster are only distinguished by their number of points and their geometry. In Eqn. (6), the $\Theta_n(\sigma)$ are given by:

$$\Theta_n(\sigma) = \sum_{\alpha \in n} \Phi_\alpha(\sigma)\,. \tag{7}$$

In view of the orthogonality of the $\Phi_\alpha(\sigma)$, we also have:

$$\frac{1}{2^N} \sum_\sigma \Theta_n(\sigma)\, \Theta_m(\sigma) = z_n\, N\, \delta_{n,m}\,, \tag{8}$$

where $z_n N$ is the total number of n-type clusters in the crystal.

The most common applications of Eqn. (6) are for the cluster expansion of expectation values of functions of configurations, such as the average of the configurational energy. With the notation $\xi_n = <\Phi_\alpha(\sigma)>$ for the expectation value of the characteristic functions, where α is any cluster belonging to the equivalent set n, we obtain:

$$\bar{F} = <F(\sigma)> = N \sum_{n=0}^{N} z_n\, F_n\, \xi_n\,. \tag{9}$$

The usefulness of this cluster expansion rests on the fast convergence of the projections F_n. In Section 3, the cluster expansion given by Eqn. (9) is used to obtain the renormalized contributions to the configurational energy arising from chemical interactions, local volume relaxations and vibrational modes.

2.2 Total Energies

As mentioned, the development of local-density functional theory [7] has been instrumental in our ability to calculate the total energy of ordered compounds from the knowledge of their electronic structure. These calculations, which use only atomic numbers as input, correctly reproduce 0 K ground state properties of the elements and of ordered compounds.

A convenient representation of the calculated total energy curves for a given compound is provided by a Morse function of the form [27]:

$$E(r) = A - 2C\, e^{-\lambda (r - r_0)} + C\, e^{-2\lambda (r - r_0)}, \tag{10}$$

where E(r) is the calculated electronic total energy of the rigid lattice and A, C, λ, and r_0 are fitting parameters. Here, the variable r, is the Wigner-Seitz atomic radius related to the volume per atom by the relation $\Omega=(4\pi/3)\, r^3$. For the compounds, r is the *effective* Wigner-Seitz radius obtained from the average of the constituent atomic volumes. It follows from Eqn. (10) that r_0 is the Wigner-Seitz radius corresponding to the minimum in the binding curve and that C is the cohesive energy of the rigid lattice.

The results of the electronic structure calculations can be extended to include the vibrational modes of configurationally ordered systems, i.e. pure elements and ordered compounds, by means of a Debye-Grüneisen analysis [27] of the calculated binding energies [26]. In particular, the analysis yields theoretical bulk moduli, Debye temperatures, and Grüneisen constants from which the vibrational free energies of pure metals and hypothetical, chemically ordered compounds, are obtained. For simple elemental systems [27] and alloys [26], the theoretical thermal properties are in reasonable agreement with experiment.

In terms of properties of the rigid lattice, the vibrational free energy, F(r,T), is given by,

$$F(r,T) = \frac{9}{8} k_B \Theta + E(r) - k_B T\, [\, D(\Theta/T) - 3 \ln(1 - e^{-\Theta/T})\,], \tag{11}$$

where k_B is Boltzmann's constant, D(x) is the Debye function, E(r) is the electronic binding energy, and where the volume dependence of the Debye temperature, Θ, is given by:

$$\Theta = \Theta_0\, (r_0/r)^{3\gamma}, \tag{12}$$

with Θ_0 the Debye temperature corresponding to r_0, and with γ the Grüneisen constant. For the ordered compounds, the free energy given by Eqn. (11) represents the volume and temperature dependent binding energy in the absence of configurational disorder.

2.3 Effective Interactions

The cluster expansion developed in Section 2.1, and in particular Eqn. (9), may be applied to the energy of a set of ordered compounds, for which the correlation functions are known *a-priori*, in order to obtain effective chemical interactions applicable to the disordered system. Here we apply the procedure to the vibrational free energies of the ordered compounds calculated in the Debye-Grüneisen approximation. [26] In practice, the applicability of the method depends upon the convergence of the cluster expansion for relatively small clusters, and on the availability of a set of total energies for which the required inversion of Eqn. (9) is defined. At present, general selection criteria for clusters giving a converged cluster expansion are not available. Thus, a maximum interaction range is assumed *a-priori*, much as is done with the correlation range in the CVM. The accuracy of the cluster expansion can be ascertained by comparing the total energy of compounds not included in the inversion procedure, with the values obtained using the assumed cluster expansion [23,24]. It should be noted that for simple cases such as the fcc lattice in the tetrahedron approximation, the inversion of Eqn. (9) is straightforward, requiring the calculation of total energies for only five high symmetry structures. For larger cluster approximations, this inversion is not immediately apparent. It can be shown, however, that within a given maximum cluster approximation, there is always a natural set of relevant structures, given by the vertices of a convex configurational polyhedron, for which the inversion of Eqn. (9) is unique [28].

Defining the correlation functions $\xi_{k,n}$, where k labels a set of ordered compounds and n labels the interactions included in the cluster expansion, the vibrational free energy (per atom) for each of the ordered structures, $F_k(r,T)$, takes the form:

$$F_k(r,T) = \sum_{n=0}^{m} z_n V_n(r,T) \xi_{k,n} \quad , \tag{13}$$

where $V_n(r,T)$ are the volume and temperature dependent effective interactions. By properly choosing the set of ordered compounds (k=0,1, ... m) and of the cluster interactions (n=0,1, ... m), Eqn. (13) can be inverted. At a fixed Wigner-Seitz radius and temperature, we have:

$$V_n(r,T) = \sum_{k=0}^{m} \omega_{n,k} F_k(r,T) \quad , \tag{14}$$

where $\omega_{n,k}$ are obtained by inversion of the matrix with elements $z_n \xi_{k,n}$.

Once configurational effects are included, a total free energy functional that depends on the configurational variables ξ_n and volume (or r). Volume relaxation can then incorporated globally by minimizing this functional with respect to r, in addition to the usual minimization with respect to the correlation functions ξ_n.

2.4 Configurational Entropy

The CVM, originally proposed by Kikuchi [2] in 1951, is formulated here in terms of a cluster expansion of the configurational entropy [16]. For a given probability distribution $X(\sigma)$, the configurational entropy is given exactly by:

$$S = -k_B \sum_{\sigma} X(\sigma) \ln X(\sigma) \quad , \tag{15}$$

where the sum is carried over all 2^N configurations in the crystal. A tractable representation of Eqn. (15) may be obtained by considering an infinite series of clusters with entropies defined by:

$$S_\alpha = -k_B \sum_{\sigma_\alpha} X_\alpha(\sigma_\alpha) \ln X_\alpha(\sigma_\alpha), \tag{16}$$

where the cluster probability distribution $X_\alpha(\sigma_\alpha)$ is given by the sum of the $X(\sigma)$ over all configurational variables σ_i outside cluster α. This series of cluster entropies trivially converges to the exact configurational entropy as the size of the cluster α increases to include all points in the crystal. Using an exact Möbius transformation, we may also write the cluster entropies, S_α, in terms of a set of *irreducible* cluster contributions, $\hat{S}_\alpha$, as: [16]

$$S_\alpha = \sum_{\beta \in \alpha} \hat{S}_\beta \quad , \tag{17}$$

where the sum runs over all the subclusters of α, including α, and excludes the empty cluster.

The key approximation made in the CVM consists of neglecting the irreducible entropy contributions $\hat{S}_\alpha$ for clusters larger than a given maximum cluster. This closure condition allow us to express the total configurational entropy, S, in terms of a finite sum of irreducible contributions. Using the space group symmetry of the crystal, Eqn. (17) for $\alpha \to N$ becomes [16]:

$$S = N \sum_{n=1}^{m} z_n \hat{S}_n \tag{18}$$

or, in terms of the cluster entropies:

$$S = N \sum_{n=1}^{m} z_n a_n S_n = -N k_B \sum_{n=1}^{m} z_n a_n \sum_{\sigma_n} X_n(\sigma_n) \ln X_n(\sigma_n) \quad , \tag{19}$$

where, as defined previously, n labels inequivalent clusters and m labels the maximum cluster. The coefficients a_n, obtained by inverting Eqn. (17), are given by: [16]

$$\sum_{\beta \in \alpha}{}' a_\beta = 1 \quad , \tag{20}$$

where the equation is valid for each subcluster α of the maximum cluster, and where the sum runs over all subclusters β of the maximum clusters that contain or equal α.

The total free energy functional (per atom) of the disordered alloy is given by:

$$F_{tot} = \sum_{n=0}^{m} z_n V_n(r,T)\, \xi_n + k_B T \sum_{n=1}^{m} z_n a_n \sum_{\sigma_n} X_n(\sigma_n) \ln X_n(\sigma_n) \,. \tag{21}$$

Here the effective interactions $V_n(r,T)$ are given by Eqn. (14). Using the cluster expansion described in Section 2.1, the cluster probability distributions can be written in terms of the multisite correlation functions [16]:

$$X_n(\sigma_n) = \frac{1}{2^n} [\, 1 + \sum_{n=1}^{m} \Theta_n(\sigma_n)\, \xi_n \,] \quad , \tag{22}$$

where the $\Theta_n(\sigma_n)$ are the characteristic functions defined as sums of products of the configurational variables σ_i for lattice sites i belonging to cluster n (see Eqns. (2) and (7)). Thus the free energy functional given by Eqn. (21) is function only of the Wigner-Seitz radius r and of the correlation functions ξ_n. At a given temperature and concentration, the latter being given by the point correlation ξ_1, the equilibrium free energy is obtained by minimizing the free energy functional with respect to the remaining correlation functions, ξ_n and the Wigner-Seitz radius r.

3. The Ru-Zr-Nb System

Total-energies as a function of atomic volume were calculated for compounds based on the hcp and bcc structures for Zr-Nb, Nb-Ru, and Zr-Ru. For the bcc and hcp based structures we used, respectively, nine and six intermetallic compounds. The respective structures are shown in Fig. 1. Opened and filled circles correspond to either type of atom, and different sizes to different locations along the [001] direction, projected onto the (001) plane.

The LMTO-ASA ground-state energies of selected compounds in Zr-Nb at ground-state are shown in Fig. 2. The effective interactions obtained using Eqn. (14)

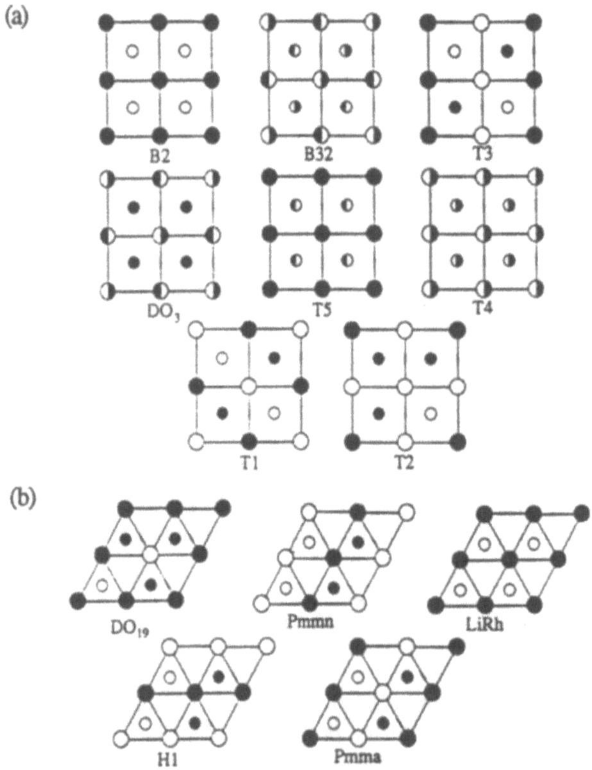

Figure 1: Ordered compounds used in the total energy calculations for: (a) bcc-based structures and (b) hcp-based structures.

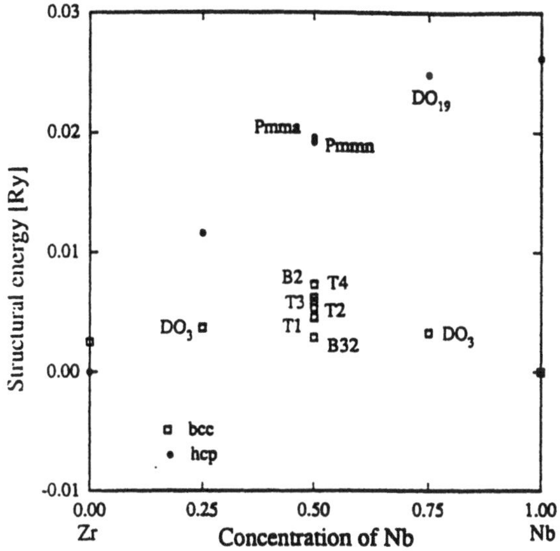

Figure 2: Calculated energies of formation for the Zr-Nb system.

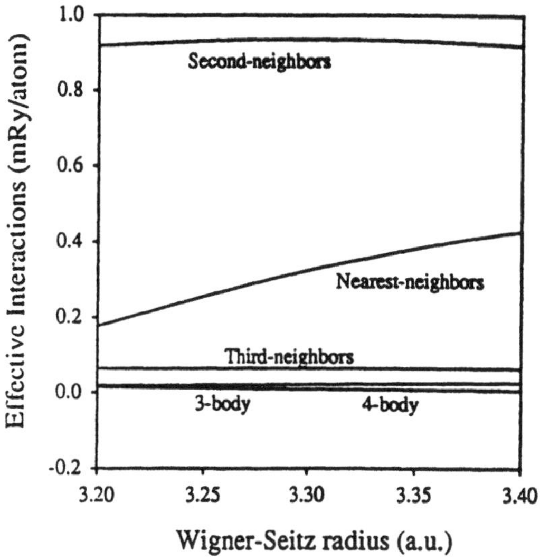

Figure 3: Effective chemical interactions as a function of the Wigner-Seitz radius for the Zr-Nb system.

are displayed in Fig. 3. The results shown in Fig. 3 suggest that the cluster expansion converges relatively quickly with the range and the complexity of the interactions (pairs vs. many-body). The convergence of the cluster expansion was also judged by comparing ground-state energies obtained from electronic structure calculations for compounds outside of the basis set, with energies predicted by the

Table 1: Energies of formation for the four tetragonal ZrNb structures calculated using the LMTO method and the errors of the cluster expansion including up to second and up to third neighbors. For the expansion including up to third neighbors errors for the three possible choices of basis are given. Energies are in mRy/atom.

Structure	ΔE (LMTO)	Error in Cluster Expansion 2nd	3rd		
ZrNb(T1)	3.3	0.15	0.00	-0.37	-0.15
ZrNb(T2)	4.6	0.39	0.28	0.00	0.17
ZrNb(T3)	4.1	0.30	0.15	-0.22	0.00
ZrNb(T4)	5.0	1.23	1.09	0.71	0.93

cluster expansion itself. Such a comparison for the bcc-based structures of the Zr-Nb system is shown in Table 1.

Using a basis set of six compounds, that includes the pure elements with the tungsten structure, the Zr_3Nb and $ZrNb_3$ with the DO_3 structure, and ZrNb with the B2 and the B32 structures, allows a cluster expansion of up to second-nearest neighbors (including three- and four-body terms). The basis was augmented with a simple tetragonal structure, T1 (see Fig. 1) in order to include third-nearest neighbors pair interactions. As seen from Table 1, a cluster expansion of up to third-nearest neighbors appears to be well converged since it allows prediction of the energies of the additional tetragonal structures, T2, T3, and T4 with an error of less than 1 mRy.

The finite temperature free energy calculations were performed using the CVM in the tetrahedron-octahedron approximation for the hcp alloys, which is consistent with the second-nearest neighbor interactions included in the Hamiltonian. For the bcc structures the largest clusters used were the bcc octahedron plus the nine points bcc unit cell, which explicitly includes up to third-nearest neighbor pairs. Thus, the CVM approximations for both the bcc and hcp structures are consistent with the range of interactions used in the cluster expansion of the energy.

The resulting phase diagram is shown in Fig. 4. The top of the predicted miscibility gap is at 1248 K and a concentration of 62% Nb, 1.3% below the measured 1261 K at 61% Nb [29]. The relationship of the bulk modulus to the Debye temperature for the hexagonal structures was adjusted to reproduce the hcp to bcc transition temperature. We emphasize however that this *ad hoc* treatment gives a Debye temperature (301 K) that is approximately 3% higher than the experimental value (291 K). With the adjusted Debye temperature the predicted monotectoid equilibrium temperature (855 K) is within 6.5% of the measured temperature (893 K) [29].

Few experimental data for the Ru-Nb system are published. Two unresolved issues of principal interest are (i) the crystal structure of the compound Ru_3Nb and (ii) the extent of the Nb-rich bcc phase and its possible phase change at approximately 40% Ru.

A high-pressure $L1_2$ phase is reported for Ru_3Nb [30] as is an unspecified hexagonal phase [31]. Using the cluster expansion method described above, a search of all possible ground-state configurations for the hcp lattice was conducted using the LMTO results for a basis set of six structures. This expansion includes up to second-nearest neighbors. The results of this search imply that of all structures with

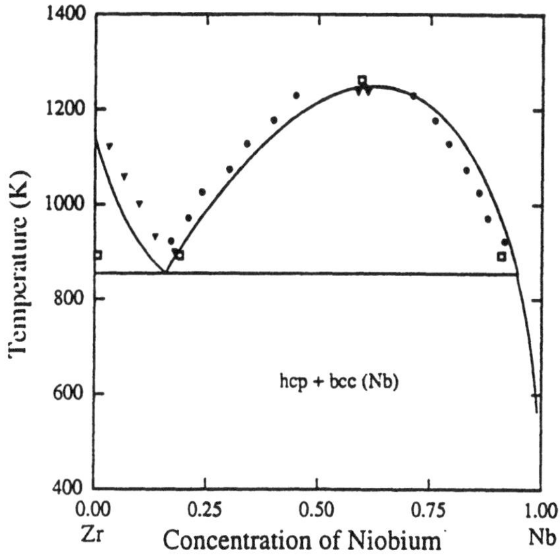

Figure 4: Calculated phase diagram for Zr-Nb.

16 or fewer atoms per unit cell, the DO_{19} structure is the most stable hcp-based compound. Furthermore, LMTO total energy calculations using a c/a ratio corresponding to that reported for Ru_3Nb (c/a = 1.6), render the DO_{19} structure more stable than fcc and bcc-based compounds of the same stoichiometry (i.e. the $L1_2$ and $D0_3$ structures, respectively). Thus, the results of our electronic structure calculations identify as DO_{19} the hexagonal Ru_3Nb phase reported experimentally and, at the same time, underscore the fact that the $L1_2$ phase is metastable with an energy very close to that of the stable DO_{19} phase.

Another aspect of interest in the Ru-Nb system is the report by Hurley and Brophy [31] of a tetragonal distortion at 37% Ru. These authors, however, did not include this phase in their proposed phase diagram, which posits a single bcc phase field in equilibrium with the compound Ru_3Nb.

In order to elucidate the stability and nature of the tetragonality observed by Hurley and Brophy, we performed LMTO calculations for NbRu in both the B2 and $L1_0$ structures. Figure 5 shows that the $L1_0$ structure should be stable at an approximate c/a ratio of 1.122 which matches well with x-ray diffraction data from Hurley and Brophy. Although not investigated here, we note that Das *et al* [32] reported a transformation from a tetragonal to an orthorhombic structure at approximately 46% Ru. Their measured c/a and b/a ratios at 51.1% Ru are, 1.114 and 1.137, respectively.

The delicate competition between different ordered compounds present in Ru-Nb is not observed in the Ru-Zr system. In fact, the experimental phase diagram for this system is well characterized and shows the presence of only a very stable B2 intermetallic compound at 1:1 stoichiometry. This observation is in qualitative agreement with the ground-state electronic energies of formation for the Zr-Ru system shown in Fig. 6. Furthermore, we obtain excellent quantitative agreement between the calculated energy of formation, ΔE^{LMTO}= -135.9 kJ mol^{-1} and the experimental value, ΔE^{exp}=-137.3 ± 6.8 kJ mol^{-1} [33].

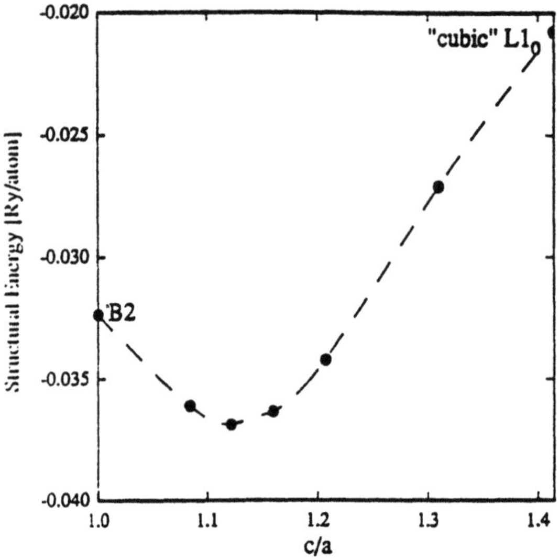

Figure 5: Calculated structural energies for the RuNb ($L1_0$) structure as a function of the c/a ratio.

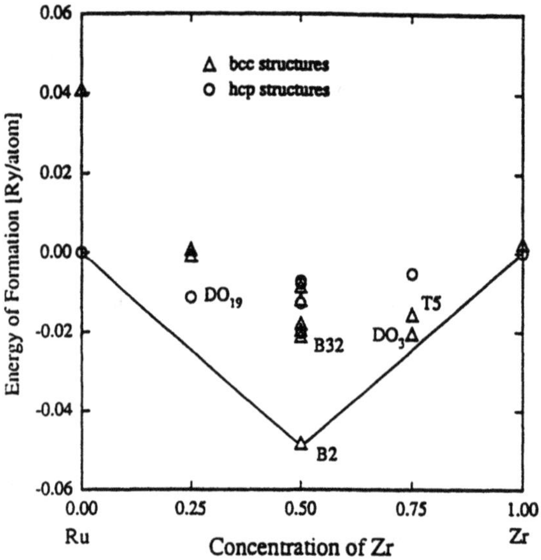

Figure 6: Calculated energies of formation for the Ru-Zr system.

4. Conclusions

The calculation of phase diagrams from the knowledge of the electronic structure of compounds has, over the last few years, emerged as a potentially useful tool in alloy design. We have attempted to demonstrate that the basic theoretical principles, as well as the limitations of the method, are presently well understood. As an example, the zirconium-niobium phase diagram has been calculated with a high degree of accuracy. Subtle aspects of phase stability in the ruthenium-niobium system have been addressed using the electronic structure and calculated total energies. The accuracy with which thermodynamic properties such as solubility, formation energy, and Debye temperatures can be calculated without empirical input underscores the potential usefulness of the approach as an additional tool in alloy development.

Acknowledgments: This work was supported in part by NSF Grant No. DMR-91-14646, the Office of Naval Research under ONR grant N00014-90-J-1185 and by a Grant for International Joint Research Project from the NEDO, Japan.

References

1. J. M. Sanchez and D. de Fontaine, Phys. Rev. B **17**, 2926 (1978).
2. R. Kikuchi, J.M. Sanchez, D. de Fontaine, and H. Yamauchi, Acta Metall. **28**, 651 (1980).
3. C. Sigli and J.M. Sanchez, Acta Metall. **33**, 1097 (1985).
4. J.M. Sanchez, J.R. Barefoot, R.N. Jarret, and J.K. Tien, Acta Metall. **32**, 1519 (1984).
5. C.E. Dahmani, M.C. Cadeville, J.M. Sanchez, and J.L. Moran-Lopez, Phys. Rev. Lett. **55**, 1208 (1985).
6. R. Kikuchi, Phys. Rev. **81**, 988 (1951); J. Chem. Phys. **60**, 1071 (1974).
7. W. Kohn and L. J. Sham, Phys. Rev. **140**, A1133 (1965); P. Hohenberg and W. Kohn, ibid. **136**, B864 (1964).
8. V. L. Moruzzi, J. F. Janak and A. R. Williams, "Calculated Electronic Properties of Metals" (Pergamon, New York 1978).
9. O.K. Andersen, O. Jepsen, and D. Glötzel, "Highlights of Condensed Mater Theory", Proceedings of the International School of Physics Enrico Fermi, North-Holland, Amsterdam, 1985.
10. M.T. Yin and M.L. Cohen, Phys. Rev. Lett **45**, 1004 (1980).
11. A.R. Wiliams, C.D. Gelatt, and V.L. Moruzzi, Phys. Rev. Lett. **44**, 429 (1980).
12. C.D. Gelatt, A.R. Wiliams, and V.L. Moruzzi, Phys. Rev. B **27**, 2005 (1985).
13. J.S. Faulkner, Prog. Mater. Sci. 27, pp. 1-187, 1982 (Pergamon Press).
14. H. Winter and G.M. Stocks, Phys. Rev. B **27**, 882 (1982).
15. H. Winter, P.J. Durham, and G.M. Stocks, J. Phys. F **14**, 1047 (1984).
16. J.M. Sanchez F. Ducastelle and D. Gratias, Physica **128A**, 334 (1984).
17. J.W.D. Connolly and A.R. Williams, Phys. Rev. B **27**, 5169 (1983).
18. K. Terakura, T. Oguchi, T. Mohri, and K. Watanabe, Phys. Rev. B **35**, 2169 (1987).
19. A.A. Mbaye, L.G. Ferreira, and A. Zunger, Phys. Rev. Lett. **58**, 49 (1987).
20. A.E. Carlsson and J.M. Sanchez, Solid State Comm. **65**, 527 (1988).

21. T. Mohri, K. Terakura, T. Oguchi and K. Watanabe, Acta Metall. **36**, 547 (1988).
22. A. Zunger, S.-H. Wei, A.A. Mbaye and G.L. Ferreira, Acta Metall. **36**, 2239 (1988).
23. S. Takizawa, K. Terakura and T. Mohri, Phys. Rev. B **39**, 5792 (1989).
24. L.G. Ferreira, S.-H. Wei and A. Zunger, Phys. Rev. B **40**, 3197 (1989); ibid. **41**, 8240 (1990).
25. M. Sluiter, D. de Fontaine X.Q. Guo R. Podloucky and A.J. Freeman, Physical Rev. B **42**, 10460 (1990).
26. J.M. Sanchez, J.P. Stark and V.L. Moruzzi, Phys. Rev. B **44**, 5411 (1991).

Phase Diagram for Long Period Stacking Variants

M. Kaburagi[1] *and D. de Fontaine*[2]

[1]Department of Informatics, Faculty of Cross-Cultural Studies, Kobe University, Tsurukabuto, Kobe 657, Japan

[2]Department of Materials Science and Mineral Engineering, University of California, Berkeley, CA 94720, USA

Abstract. Stability of long period stacking variants in the hexagonal cell is discussed by determining the structure of the lowest energy solutions of a pairwise interaction model. The interactions are extended to several neighbor layers in the model. Typically observed stacking variants, such as those observed in the Laves alloys and the transition-metal carbides, are obtained as the ground states of the model. Several examples are discussed in connection with the present results to show that an experimental observation of long period stacking variants yields a certain amount of information about the interactions. The possibility of finding new structures is also discussed.

1. Introduction

Recently much attention has been paid to one dimensional long period structures (LPS) [1]. Among these LPS, the anti-phase boundary (APB) based family, such as Cu-Pd alloy, is the one which has been studied most extensively, either experimentally [2] or theoretically [3]. Another example of one dimensional LPS is the close-packed (CP) based family. The CP based family belongs to the category of crystals in which an arrangement of atoms can be regarded as a one dimensional regular stacking of three types of layers (or layered compounds), namely A, B, C. The sequence of the stacking is conveniently expressed by the Yagodzinski notations ('ch' notation) in the case where two adjacent layers are never of the same type. Typical examples are the close-packed structures hcp (AB or h), dhcp (ABAC or ch) and fcc (ABC or c). Besides these, the LPS, such as Sm-type (ABABCBCAC or chh), are also observed in many systems. More than a few systems undergo transformations of the stacking sequence when they are subject to change of external conditions, such as pressure and alloy composition. For example, the pseudo-binary Laves phase $Mg(Ni_{1-x}Cu_x)_2$ has a stacking sequence ch, cch or c depending upon the alloy composition. In general, the Mg based Laves alloys have a tendency to transform according to ch→cch→c→chh→h as the electron-atom ratio e/a increases [4]. In this paper, we also discuss some

Springer Series in Solid-State Sciences, Volume 114
Interatomic Potential and Structural Stability Eds.: K. Terakura and H. Akai

other examples: the interstitial O-atoms in ZrO_x ($x \simeq \frac{1}{3}$) [5], metal atoms in transition-metal carbides MC_x [6], rare earth metals [7] and hexagonal $Co_{1-x}Fe_x$ [8]. Experimental studies of these examples have been carried out extensively. Table I lists the results of the studies.

On the other hand, theoretical studies of these examples have been done by several authors [9-12]: Kitano et al. have attempted to reproduce the transformation of the Mg based Laves alloy by assuming pairwise interactions between layers [10]. Bruinsma and Zangwill have analyzed the same transformation from an electronic point of view in terms of c-stacking-fault (c-fault) in h-sequence, although they neglected multi-body interactions between c-faults [11-12].

Table I. Examples of long period stacking variants. For the materials marked by †, other variants are also observed.

a) Structure versus e/a for psudobinary $Mg(X_xY_{1-x})_2$ Laves phase alloys: X,Y=Ni,Cu,Zn,Al,Ag [4].

e/a : 0.67 ∼ 0.97 ∼ 1.07 ∼ 1.80 ∼ 1.93[†] ∼ 1.95 ∼ 1.96[†] ∼ 1.98 ∼ 2.00
ch → cch → c → ch → chchh → chh → chhh → h → h

b) Structure of Interstitial O-atoms in hcp Zr: $ZrO_x(x \simeq \frac{1}{3})$[5].

x : 0.30 ∼ 0.31[†] ∼ 0.32 ∼ 0.33[†] ∼ 0.34 ∼
c → ccchh → cchh → cchhh → h

c) Structure of Metal atoms in transition-metal carbides MC_x ($x = \frac{1}{2} \sim 1$): M=V,Ta,Nb,Mo[6].

x :	$\frac{1}{2}$ ∼ $\frac{2}{3}$ ∼ $\frac{3}{4}$ ∼ $\frac{5}{6}$ ∼ 1
V, Ta	h → cchh → c → c
Nb	h → chh → cchh → c → c
Mo	h → cch → c

d) Structure versus density for rare earth metals[7].

increasing density →
h → chh → ch → c

e) Structure of $Co_{1-x}Fe_x$ alloys [8].

x : 0.00 ∼ 0.01 ∼ 0.04 ∼
h → ch → c →

The purpose of the present paper is to give a unified version of the above ordering phenomena from a phenomenological point of view. We first introduce the model in which the layers interact pairwise with each other in terms of an ABC-expression. After transforming the Hamiltonian to that in the ch-expression, we have determined the ground state of the model. On the basis of the phase diagram obtained from the ground state analysis, we discuss the constraints on the interactions which stabilize long period structures. The interaction regime in each system is also discussed in connection with the observed structures. The rest of this paper is organized as follows. The transformation of the Hamiltonian in ABC-expression to that in ch-expression, which is useful for the ground states analysis, is given in the next section (§2) for arbitrary range of the interactions. In §3, we determine the ground state of the model for the case where the range of the interactions is extended to 4th neighbor layer. We show in §4 that experimental observation of LPS yields a certain amount of information about the interactions.

2. Pairwise Interaction Model

We start with the model in which the interaction energy between k-th neighbor pair of the same (different) type of layer is given by U_k (U'_k). Because two adjacent layers are never of the same type, $U_1 = \infty$. Denoting the type of i-th layer by q_i (=A, B, or C), we obtain an expression of the Hamiltonian for the present system:

$$\begin{aligned} \mathcal{H} &= \sum_i \sum_k \left[U_k \cdot \delta(q_i, q_{i+k}) + U'_k (1 - \delta(q_i, q_{i+k})) \right] , \\ &= \sum_i \sum_k u_k \cdot \delta(q_i, q_{i+k}) + const , \end{aligned} \tag{1}$$

where $\delta(\cdots,\cdots)$ is the Kronecker's delta and $u_k = (U_k - U'_k)$. It should be noted that the Hamiltonian (1) is equivalent to the one dimensional three state Potts Model with extended range interactions [13]. The ground state of the Hamiltonian (1) can be, in principle, determined by the method of geometrical inequalities [14,15]. In the present case, however, it is preferable to express the Hamiltonian in terms of the 'ch' notation, because each layer in this expression can have one of only two states, c or h, rather than three and therefore we can reduce the maximum repeating unit to be considered. This section is devoted to the derivation of the energy expression in terms of the variables associated with the states c or h for arbitrary range of the interactions.

In order to obtain the energy expression, we first introduce the variable $z(i)$ for $i - (i+1)$ 'bond' associated with the rotation of the 'state' through the bond as

$$z(i) = \begin{cases} 1 & \text{for } (q_i, q_{i+1}) = \text{(A,B), (B,C), or (C,A),} \\ -1 & \text{for } (q_i, q_{i+1}) = \text{(B,A), (C,B), or (A,C).} \end{cases} \tag{2}$$

With use of z's, we can specify the relation between the type of i-th layer and that of $(i+k)$-th layer, i.e. the relation between q_i and q_{i+k} as

$$\begin{cases} q_i = q_{i+k} & \text{for } [z(i)+z(i+2)+\cdots+z(i+k-1)] = 0 \quad [mod\, 3], \\ q_i \neq q_{i+k} & \text{for } [z(i)+z(i+2)+\cdots+z(i+k-1)] = \pm 1\, [mod\, 3]. \end{cases} \tag{3}$$

The expression of $\delta(q_i, q_{i+k})$ is, therefore, given by

$$\delta(q_i, q_{i+k}) = \frac{1}{3}\left[2\cos\{\frac{2\pi}{3}(z(i)+z(i+1)+\cdots+z(i+k-1))\}+1\right], \tag{4}$$

and the Hamiltonian, apart from unimportant constant, is given by

$$\mathcal{H} = \sum_k \frac{2u_k}{3}\left[\sum_i \cos\{\frac{2\pi}{3}(z(i)+z(i+2)+\cdots+z(i+k-1))\}\right]. \tag{5}$$

The term $\sum_i \cos\{\frac{2\pi}{3}(z(i)+z(i+1)+\cdots+z(i+k-1))\}$ can be expanded as

$$\begin{aligned}\sum_i \cos\{\frac{2\pi}{3}(z(i)+z(i+1)+\cdots+z(i+k-1))=(-\frac{1}{2})^k\Big[&const.\\ &-3\sum_{0<k_1}(k-k_1)\{\sum_i z(i)\cdot z(i+k_1)\}\\ &+3^2\sum_{2<k_3}(k-k_3)\sum_{0<k_1<}\sum_{k_2<k_3}\{\sum_i z(i)\cdot z(i+k_1)\cdot z(i+k_2)\cdot z(i+k_3)\}\\ &+\cdots\cdots\\ &+(-3)^m\sum_{2m-2<k_{2m-1}}(k-k_{2m-1})\sum_{0<k_1<}\sum_{k_2<\cdots}\cdots\sum_{k_{2m-2}<k_{2m-1}}\\ &\{\sum_i z(i)\cdot z(i+k_1)\cdots z(i+k_{2m-1})\}\\ &+\cdots\cdots\Big],\end{aligned} \tag{6}$$

where $m=1,2,\cdots,[\frac{k}{2}]$ ($[\frac{k}{2}]$ is the integer part of $\frac{k}{2}$). By use of eq.(4), the Hamiltonian is expanded as

$$\begin{aligned}\mathcal{H} = &\sum_{0<k_1}\phi_2(k_1)\{\sum_i z(i)\cdot z(i+k_1)\}\\ &+\sum_{2<k_3}\phi_4(k_3)\sum_{0<k_1<}\sum_{k_2<k_3}\{\sum_i z(i)\cdot z(i+k_1)\cdot z(i+k_2)\cdot z(i+k_3)\}\\ &+\cdots\cdots\\ &+\sum_{2m-2<k_{2m-1}}\phi_{2m}(k_{2m-1})\sum_{0<k_1<}\sum_{k_2<\cdots}\cdots\sum_{k_{2m-2}<k_{2m-1}}\end{aligned}$$

$$\{\sum_i z(i)\cdot z(i+k_1)\cdots z(i+k_{2m-1})\}$$
$$+\cdots\cdots\cdots\cdots\cdots\cdots\cdots\cdots\cdots\cdots\quad , \tag{7}$$

where

$$\phi_2(k)=-2\sum_{0<j}(-\frac{1}{2})^{k+j} j\cdot u_{k+j},$$
$$\phi_4(k)=+2\cdot 3\sum_{0<j}(-\frac{1}{2})^{k+j} j\cdot u_{k+j},$$
$$\phi_{2m}(k)=(-1)^m\cdot 2\cdot 3^{m-1}\sum_{0<j}(-\frac{1}{2})^{k+j} j\cdot u_{k+j}. \tag{8}$$

It is easy to obtain the ch-expression of our model by introducing a new pseudo-Ising-spin variable $\sigma(i)$ defined by

$$\sigma(i)=z(i-1)\cdot z(i)\ , \tag{9}$$

where $\sigma(i)$=-1 or +1, for i-th layer=h or c, respectively. Using the relations $\sigma(i)\sigma(i+1) = z(i-1)\cdot z(i+1), \cdots$ etc., we obtain the pseudo-Ising-spin-expression for the Hamiltonian of the present system as

$$\mathcal{H} = \sum_i[\phi_2(1)\sigma(i) + \phi_2(2)\sigma(i)\sigma(i+1) + \phi_2(3)\sigma(i)\sigma(i+1)\sigma(i+2) + \cdots]$$
$$\sum_i[\phi_4(3)\sigma(i)\sigma(i+2) + \phi_4(4)\{\sigma(i)\sigma(i+1)\sigma(i+3) + \sigma(i)\sigma(i+3)$$
$$+\sigma(i)\sigma(i+2)\sigma(i+3)\} + \phi_4(5)\{\sigma(i)\sigma(i+2)\sigma(i+3)\sigma(i+4)$$
$$+\sigma(i)\sigma(i+3)\sigma(i+4) + \sigma(i)\sigma(i+4) + \sigma(i)\sigma(i+1)\sigma(i+3)\sigma(i+4)$$
$$+\sigma(i)\sigma(i+1)\sigma(i+4) + \sigma(i)\sigma(i+1)\sigma(i+2)\sigma(i+4)\}$$
$$+\cdots\cdots\cdots]+$$
$$+\sum_i[\phi_6(5)\{\sigma(i)\sigma(i+2)\sigma(i+4) + \cdots\cdots\} + \cdots\cdots\cdots\cdots]$$
$$+\cdots\cdots\cdots\cdots\cdots\cdots\cdots\cdots\cdots,$$

$$= J_1M_1 + \sum_{0<k_1} J_2(k_1)M_2(k_1) + \sum_{0<k_1}\sum_{<k_2} J_3(k_1,k_2)M_3(k_1,k_2)$$
$$+\sum_{0<k_1}\sum_{<k_2}\sum_{<k_3} J_4(k_1,k_2,k_3)M_4(k_1,k_2,k_3)$$
$$+\sum_{0<k_1}\sum_{<k_2}\sum_{<k_3}\sum_{<k_4} J_5(k_1,k_2,k_3,k_4)M_5(k_1,k_2,k_3,k_4)$$
$$+\cdots\cdots\cdots\cdots\quad , \tag{10}$$

where

$$M_m(k_1,k_2,\cdots,k_{m-1})=\sum_i \sigma(i)\sigma(i+k_1)\sigma(i+k_2)\cdots\sigma(i+k_{m-1}), \quad (11)$$

and $J_m(k_1,k_2,\cdots,k_{m-1})$ is the m-body interaction energy between the pseudo-Ising-spin's, being represented by a linear combination of $\phi_{2j}(k)$. Note that J_1 plays the role of a "magnetic field" for the pseudo-Ising-spin. A more explicit expression of $J_m(k_1,k_2,\cdots,k_{m-1})$ for the case where $u_k=0$ for $k>6$ is given as

$$\begin{aligned}
&J_1=\phi_2(1)=-\frac{1}{2}u_2+\frac{2}{4}u_3-\frac{3}{8}u_4+\frac{4}{16}u_5-\frac{5}{32}u_6,\\
&J_2(1)=\phi_2(2)=+\frac{1}{4}u_3-\frac{2}{8}u_4+\frac{3}{16}u_5-\frac{4}{32}u_6,\\
&J_2(2)=\phi_4(3)=3[+\frac{1}{8}u_4-\frac{1}{16}u_5+\frac{1}{32}u_6],\\
&J_2(3)=\phi_4(4)=3[-\frac{1}{16}u_5+\frac{2}{32}u_6],\\
&J_2(4)=\phi_4(5)=3[+\frac{1}{32}u_6],\\
&J_3(1,2)=\phi_2(3)=-\frac{1}{8}u_4+\frac{2}{16}u_5-\frac{3}{32}u_6,\\
&J_3(1,3)=J_3(2,3)=\phi_4(4)=3[-\frac{1}{16}u_5-\frac{2}{32}u_6],\\
&J_3(1,4)=J_3(3,4)=\phi_4(5)=3[+\frac{1}{32}u_6],\\
&J_3(2,4)=\phi_6(5)=9[-\frac{1}{32}u_6],\\
&J_4(1,2,3)=\phi_2(4)=+\frac{1}{16}u_5-\frac{2}{32}u_6,\\
&J_4(1,2,4)=J_4(1,3,4)=J_4(2,3,4)=\phi_4(5)=3[+\frac{1}{32}u_6],\\
&J_5(1,2,3,4)=\phi_2(5)=-\frac{1}{32}u_6. \qquad (12)
\end{aligned}$$

It should be noted that in the Hamiltonian (9), the number of states (=types) at each layer has been reduced from three (A,B,C) to two (c,h), however, multi-body interactions are needed to describe the pairwise interactions in the ABC-expression.

Another convenient ch-expression for the present system is the pseudo-lattice-gas model in which the Hamiltonian is obtained by replacing the pseudo-Ising-spin $\sigma(i)$ by $\sigma(i)=2n(i)-1$ ($n(i)$ is the c-fault-occupation-number at i-th layer). Note that $n(i)=1$ (0) for i-th layer=c (h). The resultant Hamiltonian can be expressed as

$$
\begin{aligned}
\mathcal{H}=&V_1P_1+\sum_{0<k_1}V_2(k_1)P_2(k_1)+\sum_{0<k_1}\sum_{<k_2}V_3(k_1,k_2)P_3(k_1,k_2)\\
&+\sum_{0<k_1}\sum_{<k_2}\sum_{<k_3}V_4(k_1,k_2,k_3)P_4(k_1,k_2,k_3)\\
&+\sum_{0<k_1}\sum_{<k_2}\sum_{<k_3}\sum_{<k_4}V_5(k_1,k_2,k_3,k_4)P_5(k_1,k_2,k_3,k_4)\\
&+\cdots\cdots\cdots\cdots \qquad , \qquad (13)
\end{aligned}
$$

where

$$
P_m(k_1,k_2,\cdots,k_{m-1})=\sum_i n(i)n(i+k_1)n(i+k_2)\cdots n(i+k_{m-1}), \quad (14)
$$

and $V_m(k_1,k_2,\cdots,k_{m-1})$ is the m-body interaction between the c-faults. Note that V_1 plays role of the "chemical potential" for the c-faults. An even more explicit expression of $V_m(k_1,k_2,\cdots,k_{m-1})$ for the case where $u_k=0$ for $k>6$ is given as

$$
\begin{aligned}
&V_1=-[u_2+2u_4+3u_6],\\
&V_2(1)=[u_3+2u_5],\\
&V_2(2)=2[u_4+2u_6],\\
&V_2(3)=u_5,\\
&V_2(4)=2u_6,\\
&V_3(1,2)=-[u_4+2u_6],\\
&V_3(1,3)=V_3(2,3)=-2u_5,\\
&V_3(1,4)=V_3(3,4)=-u_6,\\
&V_3(2,4)=-4u_6,\\
&V_4(1,2,3)=u_5,\\
&V_4(1,2,4)=V_4(1,3,4)=V_4(2,3,4)=2u_6,\\
&V_5(1,2,3,4)=-u_6. \qquad (15)
\end{aligned}
$$

The pseudo-lattice-gas Hamiltonian (13) has been used by Kitano et al. [10] and by Bruinsma and Zangwill[11], although they neglected multi-body interactions. Instead, Bruinsma et al. derived the general expression of V_1 and $V_2(k)$, which is easily obtained in the present transformation, as

$$
\begin{aligned}
&V_1=-\sum_{k>0}k\cdot u_{2k},\\
&V_2(2m)=2\sum_{k>0}k\cdot u_{2k+2m},
\end{aligned}
$$

$$V_2(2m+1)=\sum_{k>0} k\cdot u_{2k+2m+1}. \tag{16}$$

They also discussed the stability of the stacking sequences by use of the modified Blandin-Friedel-Saada potential.

Starting from the pairwise interaction model in the ABC-expression, we have derived the ch-expression, as shown in either eq.(9) or eq.(12), which includes multi-body interactions.

3. Ground State Analysis

In this section, we deal with the ground state structure for the case where $u_k=0$ for $k \geq 5$. The explicit form of the Hamiltonian (13) for this case is given by

$$\begin{aligned}\mathcal{H}=\sum_i \big[&-(u_2+2u_4)n(i)+u_3 n(i)n(i+1)\\ &+2u_4 n(i)n(i+2)-u_4 n(i)n(i+1)n(i+2)\big]\ .\end{aligned} \tag{17}$$

In general, it is difficult to analyze rigorously the ground state of a system with multi-body interactions. However, in this case, we are considering a one dimensional system for which the possible states of lowest energy are determined completely by the range of the interaction, irrespective of its multiplicity[16]. In the present case, the repeating units of the possible states are c, cch, ch, cchh, chh, and h. The energies of these structures are given by

$$\begin{aligned}E(c)&=-u_2+u_3-u_4,\\ E(cch)&=\frac{1}{3}[-2u_2+u_3-2u_4],\\ E(ch)&=-\frac{1}{2}u_2,\\ E(cchh)&=\frac{1}{4}[-2u_2+u_3-4u_4],\\ E(chh)&=-\frac{1}{3}[u_2+2u_4],\\ E(h)&=0.\end{aligned} \tag{18}$$

By comparing the energies of these structures, we obtain the phase diagram in u_2-u_3 space shown in Fig.1. It is noted that many other variants are degenerate in energy on the boundary lines.

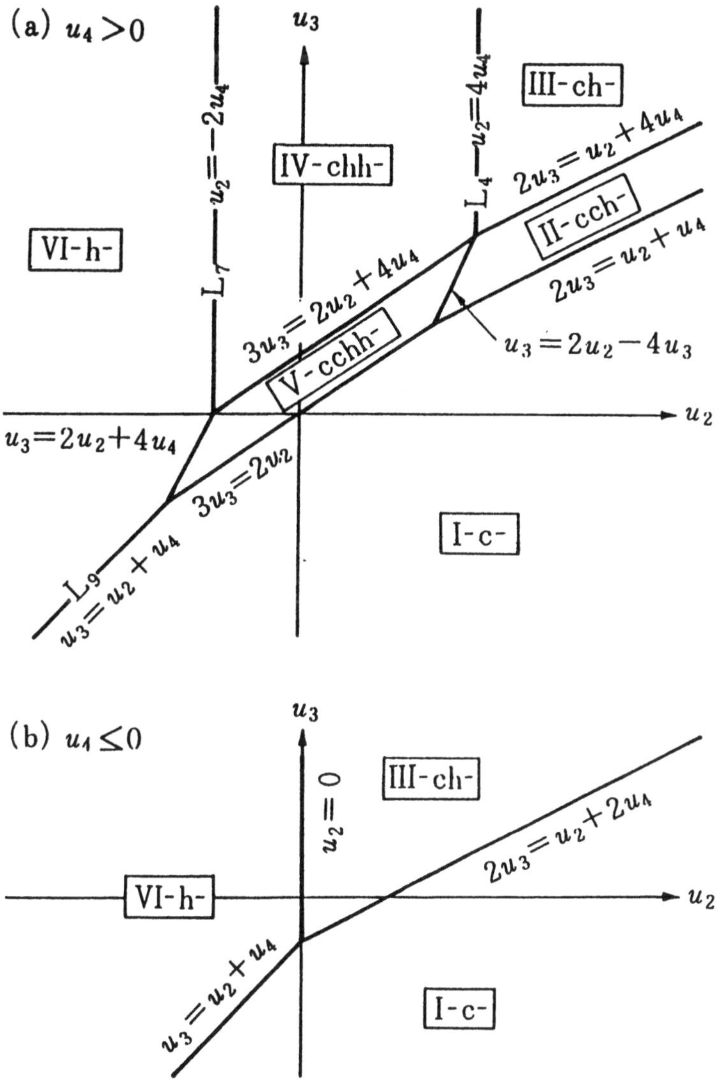

Fig. 1. Phase diagram in diagram in $u_2 - u_3$ space: (a) $u_4 > 0$, (b) $u_4 \leq 0$. The most stable stacking sequence in each region (labeled by I to VI) is shown in this phase diagram.

4. Discussion

On the basis of the phase diagram shown in Fig.1, we discuss several systems exhibiting the long period stacking variants to demonstrate that the present analysis facilitates the systematic understanding of the phenomena mentioned in §1. It may be natural to suppose that the interaction constants of the system vary in accordance with the external conditions, such as e/a; the system undergoes the transformation of the stacking sequence when the locus of interaction constants in u_k-space passes through a boundary lines, such as L_4, L_7 in Fig.1. From this point of view, the transformations in the systems listed in Table I are interpreted as the results of the following systematic variation in the interaction constants;

(a)$Mg(X_xY_{1-x})_2$:III → II → I → L_4 → IV → L_7 → VI;
(b)ZrO_x:I → V → VI;
(c)VC_x, TaC_x, NbC_x:VI → (IV) → V → I; MoC_x:VI → II → I;
(d)Rare earth:VI → IV → III → I;
(e)$Co_{1-x}Fe_x$:I → III → VI.

Thus, the region and locus of the interaction constants can be deduced from experimental observation of stacking variants.

On the other hand, once the locus of the interaction constants is determined either experimentally or theoretically, we can predict the stabilized stackings and transformations among them by use of the phase diagram. This is another advantage of the present analysis. In the case of MoC_x, for example, it is suggested from the locus VI → II → I that there is a possibility of finding the structures 'chh' and 'cchh' within $0 < x < \frac{2}{3}$. A similar argument as this may be applied to other systems. The investigation of these possibilities may be the subject of future experimental research.

Acknowledgements

The present work has been supported in part by a Grant-in-Aid for Scientific Research on Priority Areas, "Computational Physics as a New Frontier in Condensed Matter Research", from the Ministry of Education, Science and Culture. One of the authors (M. K.) gratefully acknowledges the support of the Fujitsu Limited.

References

1. *Modulated Structure Materials*, ed. T. Tsakalakos, (Martinus Nijhoff Publisher 1984)
2. S. Takeda, J. Kulik and D. de Fontaine: J. Phys. F. Metal Phys. **18** (1988) 1387
3. G. Ceder, M. De Graef, L. Delaey, J. Kulik and D. de Fontaine, Phys. Rev. **B39** (1989) 381; G. Ceder, D. de Fontaine, H. Dreysse, D. M. Nicholson, G. M. Stocks and B. L. Gyorffy, to be published in Acta Metall.
4. Y. Komura and Y. Kitano, Acta Cryst. **B33** (1977) 2496
5. S. Yamaguchi and M. Hirabayashi, J. Appl. Cryst. **3** (1970) 319; *Order-Disorder Transformation in Alloys*, ed. H. Warlimont (Springer-Verlag, 1974)
6. C. H. deNovion and V. Maurice, J. Phys. (France) **38** (1977) C7-211; L. E. Toth, *Transition Metal Carbides and Nitrides* (Academic Press, 1971) p.29
7. D. B. McWhan et al., Phys. Rev. **154** (1967) 438
8. T. Onozuka et al., J. Phys. Soc. Jpn **37** (1974) 687
9. M. Kaburagi, AIP Conf. Proc. No.**53** *Modulated Structures-1979* p.127, ed. J. M. Cowley, J. B. Cohen, M. B. Salamon and B. J. Wuensch (AIP, 1979): Preliminary report of this study has given in this reference.
10. Y. Kitano, T. Ohba and Y. Komura, AIP Conf. Proc. No.**53** *Modulated Structures-1979* p.130, ed. J. M. Cowley, J. B. Cohen, M. B. Salamon and B. J. Wuensch (AIP, 1979)
11. R. Bruinsma and A. Zangwill, Phys. Rev. Lett. **55** (1985) 214
12. for review see, for example A. Zangwill and R. Bruinsma: Comments Cond. Mat. Phys. **13** (1987) 1; R. Bruinsma, Phase Transitions, **16/17** (1989) 275; R. Bruinsma and A. Zangwill, Phys. Rev. Lett. **55** (1985) 214
13. for review see, for example F. Y. Wu, Rev. Mod. Phys. **54** (1982) 235
14. T. Tonegawa, K. Takasaki, M. Kaburagi and J. Kanamori, J. Phys. Soc. Jpn. **54** (1985) 3326
15. T. Tonegawa, K. Takasaki and M. Kaburagi, J. Phys. Soc. Jpn. **54** (1985) 4321
16. T. Morita, J. Phys. **A7** (1978) 289 and 1618

Kinetic Study of an Alloy Phase by the Path Probability Method

T. Mohri

Department of Metallurgical Engineering, Hokkaido University, Sapporo 060, Japan

Abstract. The Path Probability Method is employed to study time evolution of short range and long range order parameters of an fcc based spin system. Non-monotonous time evolution of an octahedron cluster is found for a disordered phase, which can be interpreted in terms of kinetic path in a thermodynamic configuration space. The distinction between the nucleation-growth type ordering and the spinodal ordering is revealed from the kinetic point of view. The extension of the present study of a spin system to an alloy system with the vacancy mechanism is discussed.

1 Introduction

Once an analytical expression of a free energy functional is obtained, one can derive relevant thermodynamic quantities of an alloy system such as a phase diagram, spinodal ordering temperature [1], short range order diffuse intensity spectrum in a k-space [2] from a single formula. The Cluster Variation Method [3] (hereafter CVM) of statistical mechanics has been recognized as a powerful tool to provide an accurate free energy and has been extensively employed for the study of alloy thermodynamics. By combining with electronic structure calculations, it is now a common practice to perform first-principles thermodynamic calculations. [4, 5]

Contrary to the remarkable development of equilibrium thermodynamics, less progress has been achieved for the kinetic studies. Since one of the main strategies in designing an advanced material is an employment of a metastable state of a given material through non-equilibrium processes, the time dependent stability is quite an important subject. About a decade after Kikuchi devised the CVM, he proposed the Path Probability Method [6, 7] (hereafter PPM). Since the PPM is the natural extension of the CVM to the time domain one can expect various fruitful features inherited from the CVM. The difficulty of this method, however, is originating from the fact that the method inherently demands the enormous number of variational parameters which complicate the mathematical

Springer Series in Solid-State Sciences, Volume 114
Interatomic Potential and Structural Stability Eds.: K. Terakura and H. Akai

procedures. Then, the applicability of the PPM to an alloy system has been limited to bcc based systems [8] for which the pair approximation provides a satisfactory result. It is well known, however, that the pair approximation is not sufficient for an fcc based system and, at least, the tetrahedron approximation is required to attain a reliable result. Although the extension of the PPM formulation within the tetrahedron approximation has been under way [9, 10] fully satisfactory calculation has not been yet achieved.

Both the spin and alloy systems are equivalent in the description of an equilibrium state for the Ising model. However, the formulation of the kinetics for a spin system and an alloy system is quite different. The latter characterized by the exchange mechanism strictly conserves the concentration, while the spin system described by the flip mechanism is not a conserved one. Generally, the PPM formulation as well as the computation for the spin system demands less laborious procedure.

As an initial attempt to study ordering kinetics for an fcc based system by the PPM, the author's group has been carrying out preliminary calculations on the spin system. [11, 12, 13, 14, 15] And in order to simulate an alloy system, the particular attention has been limited to 1:1 stoichiometric composition for which the concentration of up and down spins is conserved without additional constraints. The main purpose of this brief report is to provide an overview of spin kinetics for an fcc based system performed by the author's group based on the PPM calculation. The particular focus is placed on the ordering relaxation phenomenon of a disordered solid solution and the kinetic study of nucleation-growth type ordering and spinodal ordering. The organization of the present report is as follows. In the next section, a brief outline of the theoretical background of the PPM is described. The major results are demonstrated and discussed in the third section. The extension of the present calculation for the spin system to an alloy system is briefly discussed in the final section.

2 Theoretical back ground of the PPM

As the counter part of the grand potential of the CVM, the Path Probability Function $P(t; \Delta t)$ (hereafter PPF) is defined for the PPM which explicitly incorporates time as a variable. The correlation functions or cluster probabilities are defined as independent variables of grand potential for the CVM and the grand potential is minimized with respect to the correlation functions to obtain an equilibrium state. For the PPM, on the other hand, a set of transition probability from one local configuration to another during an infinitesimal time Δt is defined as independent variables which are termed path variables. And the PPF P is described in terms of the path variables as the product of the following three terms,

$$P_1 = (\theta \Delta t)^{N(X_{1,2}+X_{2,1})} (1 - \theta \Delta t)^{N(X_{1,1}+X_{2,2})} , \tag{1}$$

$$P_2 = \exp\left(-\frac{\Delta E}{2k_B T}\right) , \tag{2}$$

and

$$P_3 = \frac{\left(\prod_{ij,kl} NY_{ij,kl}!\right)^6 (N!)}{\left(\prod_{ijkl,mnop} NW_{ijkl,mnop}!\right)^2 \left(\prod_{i,j} NX_{i,j}!\right)^5} , \tag{3}$$

where N is the total number of lattice points, θ is the spin flip probability per unit time, $X_{i,j}$, $Y_{ij,kl}$ and $W_{ijkl,mnop}$ are the path variables which describe the transition probability from one spin configuration to another designated by subscript (up spin or down spin) before and after the comma sign, on a point, pair and tetrahedron clusters, respectively, ΔE is the change of the internal energy before and after the flipping event. The first term P_1 describes a statistical average of non-correlated spin flip events over an entire lattice points, and the second term P_2 is the conventional thermal activation process. The third term P_3 characterizes the PPM. One may see the similarity with the configurational entropy term of the CVM. In the PPM, this term describes the freedom of the microscopic path from one configuration to another.

Then, the PPM claims that the most probable path of the time evolution of a system is determined so that the PPF is maximized with respect to the path variables, which corresponds to the minimization condition of the CVM. By solving the set of simultaneous equations to optimize the path variables at each time step Δt, one can obtain cluster probabilities at time $t + \Delta t$ from the knowledge of those at time t. By repeating this procedure, the time evolution or devolution of the cluster probabilities for an assigned initial condition can be pursued.

3 Results and Discussions

3.1 Order relaxation in a disordered phase

Shown in Fig.1 [12, 16] are the most stable ordered phases at the ground state for 1:1 stoichiometric composition as a function of α defined as

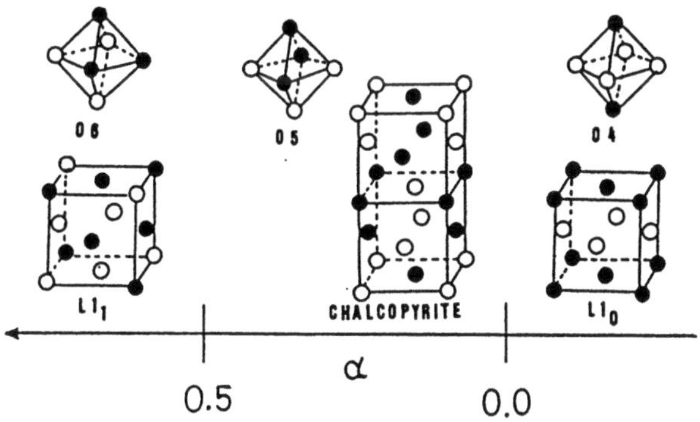

Figure 1: Most stable ordered phases and corresponding octahedron atomic arrangements as a function of α value defined as the ratio of the second to the first nearest neighbor pair interaction energies.

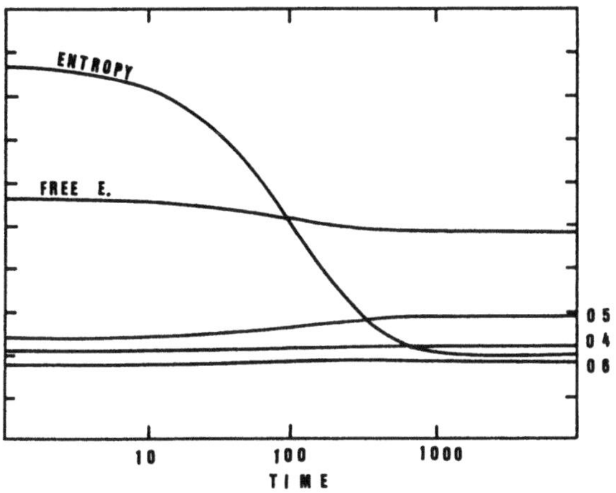

Figure 2: Time evolution of the three kinds of octahedron cluster shown in the Fig.1, free energy and the entropy for α=0.35. The time axis is in logarithmic scale and is normalized by spin flip probability θ and the vertical axis is in arbitrary unit.

the ratio of the second to the nearest neighbor pair interaction energies. The octahedron clusters shown above are the ones which represent most well the corresponding ordered phase. The time evolution of the cluster probabilities for these three clusters during an isothermal aging process at temperature $k_BT/v_{2,1}$=2.5 following the quenching operation from $k_BT/v_{2,1} = 5.0$ are calculated within the tetrahedron-octahedron approximation for α=0.35 with a fixed spin flip probability and are plotted in Fig.2. [12] Note that the time axis throughout the present study is in a logarithmic scale and is normalized with respect to the spin flip

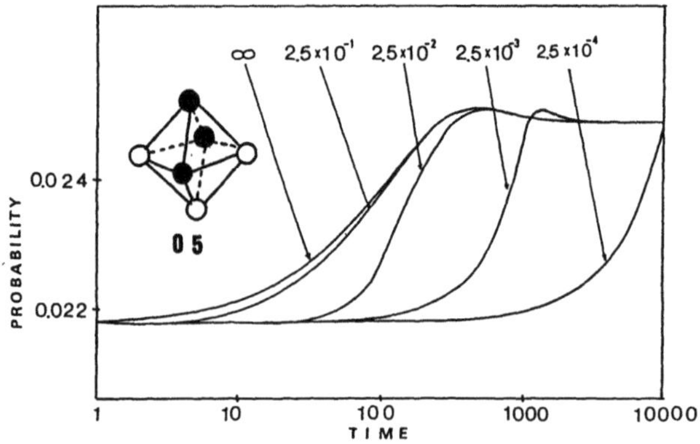

Figure 3: Time evolution of Chalcopyrite cluster at temperature 2.5 following the quenching from 5.0 with various cooling rate designated in the figure.

probability and that the temperature is, hereafter, normalized by the nearest neighbor pair interaction energy $v_{2,1}$. One sees that the chalcopyrite cluster(O5) is most stable for an entire time range, which reflects the stability at the ground state (Fig. 1). Monotonous decreases of the entropy and free energy imply that the system approaches equilibrium state with an enhancement of ordering tendency. The steady state value of each cluster probability in the long time limit is confirmed to be exactly the equilibrium one independently obtained by the CVM.

Interestingly, the time evolution of the cluster probability for the chalcopyrite cluster(O5) is not monotonous for α=0.65 as are shown in Fig. 3. [13, 14] Depending upon cooling rate a peaking is observed before reaching the final equilibrium state. The thermodynamic origin of such a non-monotonous behavior is well understood as the curvature of the kinetic path in a thermodynamic configuration space. In fact, the path is not necessarily along the steepest decent direction of the equi-free energy surface, which is due to the kinetic effects. It should be noted that the free energy of the system always decreases irrespective of the non-monotonous behavior of cluster probabilities.

3.2 Nucleation-Growth ordering and Spinodal ordering

The discussions in the previous section are limited to the order relaxation behavior in a disordered phase. In order to extend the PPM to the study of transition behavior, PPM formulation is extended to the $L1_0$ ordered phase for which sublattices are introduced. Since the distinction of the path variables between the sublattices should be explicitly considered, a

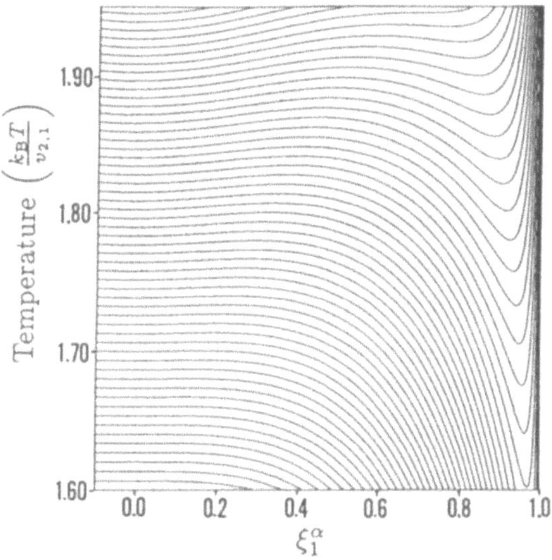

Figure 4: Free energy contour at 1:1 stoichiometry. The vertical axis is normalized temperature and the horizontal axis is the long range order parameter.

considerable mathematical burden is demanded. The present formulation is, therefore, limited to the tetrahedron approximation for which only nearest neighbor pair interaction energy is incorporated.

Among the various calculations performed, demonstrated below is the one undertaken to examine the nucleation-growth ordering (hereafter NG) and spinodal ordering (hereafter SP) from the kinetic view point. Shown in Fig. 4 is the free energy contour plot at 1:1 stoichiometric composition. The vertical axis is the temperature and the horizontal axis is point correlation function for α-sublattice which serves as a long range order parameter. Within the tetrahedron approximation of the CVM, eight configuration variables including long range order parameters are required to describe the $L1_0$ ordered phase. Among them, the number of independent variables are reduced to five for a fixed 1:1 stoichiometric composition due to a symmetry. The free energy contour in the Fig. 4 is obtained by minimizing the free energy functional under a given long range order parameter with respect to the rest of the variables for each temperature. It should be noted that the disordered phase is characterized by the vanishing long range order parameter while the finite value of the long range order parameter indicates an ordered state. One can find that the transition temperature is around 1.89, where the free energies of a disordered phase and an ordered phase become equivalent. The SP temperature is also determined to be 1.63 as the onset of vanishing of the

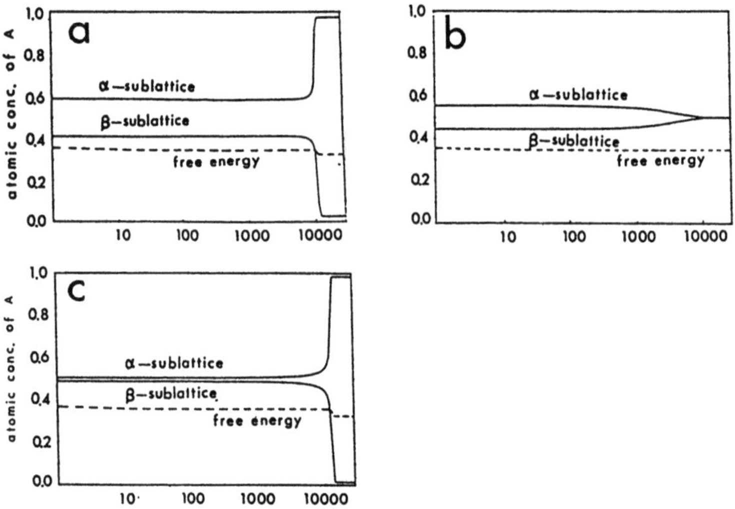

Figure 5: Time evolution of spin concentration on α and β sublattices at temperature 1.70 (Fig. 5a and 5b) and at 1.60 (Fig. 5c).

inflection point of the free energy curve. The PPM calculations are performed for two temperatures, 1.70 and 1.60 which are, respectively, above and below the SP temperature. In both cases the system is maintained at temperature 2.50 prior to the quenching operation to each temperature with an infinite cooling rate.

Shown in Figs.5a, and 5b [15] are the time evolution of the atomic concentration of A atom on both α- and β-sublattices which are regarded as the long range order parameter. In order to trigger the phase transition, the concentration fluctuation is imposed at temperature 2.50 which is indicated by the initial separation. When the imposed concentration fluctuation is large (Fig. 5a), the ordering proceeds and the $L1_0$ ordered phase is formed. In fact, the steady state values of atomic concentrations of both α- and β-sublattices in the long time limit are confirmed to be the equilibrium ones independently obtained by the CVM. While for a small fluctuation (Fig. 5b) [15], the initial fluctuation fades away and system goes back to a disordered state. These are the characteristic feature of the NG ordering which requires certain amount of fluctuation to overcome energy barrier. While for the temperature 1.60, the small amount of fluctuation can be amplified and the system is decomposed into the ordered phase as is shown in Fig. 5c. This barrierless transition characterizes the spinodal ordering reaction.

Among the various kinetic theories, the advantage of the PPM is the fact that the PPM provides the time evolution of short range order pa-

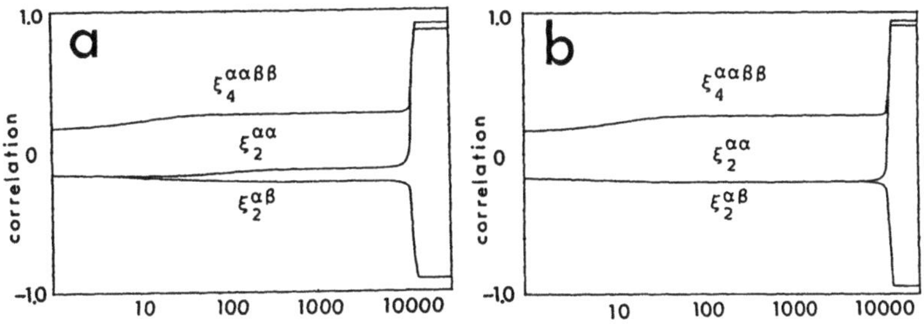

Figure 6: Time evolution of short range order parameters (tetrahedron correlation function and two types of pair correlation functions) at temperatures 1.70 (Fig. 6a) and 1.60 (Fig. 6b).

rameters in addition to the long range order parameters. Shown in Figs. 6a and 6b [15] are the two kinds of pair correlation functions and tetrahedron correlation function. For both types of ordering, one can observe that the short range order parameters start to relax before the onset of the abrupt change of long range order parameters (Figs.5 a, b and c). However, noticeable difference is the fact that the decoupling of the original lattice into two sublattices which is characterized by the separation of the two types of pair correlation functions takes place in an earlier period for the NG ordering. The thermodynamic origin of this intriguing behavior is now under study based on the kinetic path in a thermodynamic configuration space.

4 Extension to an Alloy system

The body of the present report is devoted to an fcc spin system. In this final section, we briefly discuss the extension of the PPM formula of the spin system to that of an alloy system. [9, 10] Among three factors of PPF given in eqs.(1) to (3), the extension of the third term which provides the number of equivalent path is quite straight forward. Hence the main focus is limited to the P_1 and P_2 terms which correspond to the Boltzman factor in the equilibrium thermodynamics.

We assume that the atomic motion is limited to the nearest neighbor jumping with the vacancy mechanism. Then the P_1 term for an alloy system can be given as

$$P_1 = (\theta \exp(-\beta U)\Delta t)^{C_1 \sum_{m\neq 0}\{Y_0(m0;0m)+Y_0(0m;m0)\}}, \tag{4}$$

where C_1 is a constant determined by the coordination number, θ is the vibration frequency, β is defined as $1/k_{\mathrm{B}}T$ and U is the intrinsic activation energy i.e. the saddle point energy. $Y_0(m0:0m)$ ($Y_0(0m:m0)$) is the pair path variable for an A or B atom specified by m exchanging with an adjacent vacacy which is denoted by 0.

The second term P_2 is given as

$$P_2 = \exp\left[-C_2\beta\left\{\sum_{m\neq 0}\sum_{j\neq 0}\sum_{h=1}^{4}\sum_{direction}^{g_h}\epsilon(m,j)Y_h(mj;0j)\right\}\right], \tag{5}$$

where C_2 is another constant related to the coordination number, m and j denote A or B atom and the suffix h specifies four kinds of different jumping scheme which is classified based on the symmetry of jumping-in sites [10], g_h is the degeneracy factor for each jumping direction and $\epsilon(m,j)$ is the nearest neighbor bonding energy between species m and j. It should be noted that the pair path variable in this term is limited to the one, $Y_h(mj;0j)$, describing jumping out to an adjacent vacant site, which indicates the breaking of chemical bond between m and j. Hence the P_2 term describes the total energy to break the chemical bonds in the system.

It may not be a difficult task to calculate $\epsilon(m,j)$ by electronic structure calculations. The main difficulty, however, is to evaluate the saddle point energy U which is anticipated to be sensitively dependent on local environment. We believe that the accurate calculation of U is the key to the success of first-principles kinetic investigation.

5 Acknowledgements

The author acknowledges a support by a Grant for International Research Project from the NEDO, Japan.

References

[1] D. de Fontaine, Acta metall. **23**, 553 (1975).

[2] T Mohri, J. M. Sanchez and D. de Fontaine, Acta metall. **33**, 1463 (1985).

[3] R. Kikuchi, Phys. Rev. **81**, 998 (1951).

[4] T Mohri, K. Terakura, T. Oguchi and K. Watanabe, Acta metall. **36**, 547 (1987).

[5] T .Mohri, K. Terakura, S. Takizawa and J. M. Sanchez, Acta metall. **39**, 493 (1991).

[6] R. Kikuchi, Prog. Theoret. Phys. Kyoto Suppl. **35**, 1(1966).

[7] F. Ducastelle, to be appeared in Prog. Theoret. Phys. Kyoto .

[8] H. Sato and R. Kikuchi, Acta metall. **24**, 797 (1976).

[9] R. Kikuchi, Report to NIST

[10] T. Mohri, to be published

[11] T. Mohri, Acta metall. **38**, 2445 (1990).

[12] T. Mohri and Y. Sugawara, Proc. The First International Conference and Exhibition on Computer Application to Materials Science and Engineering-CAMSE'90 Ed. by M. Doyama, NORTH-HOLLAND, 853 (1991).

[13] T. Mohri, Proc. Intn'l. Symp. Intermetallic Compounds (JIMIS-6), Ed. O. Izumi, The Japan Institute of Metals, 209 (1991).

[14] T. Mohri, in *Structural and Phase Stability of Alloys*, Ed. by J. L. Moran-Lopez et al., Plenum, 87 (1992).

[15] T. Mohri, Proc. NATO ASI in Rhode in print

[16] M. J. Richard and J. W. Cahn, Acta Metall. **19**,1263 (1966).

Part IV

Beyond the Local-Density Approximation

Correlated Wavefunction Quantum Monte-Carlo Approach to Solids

S.G. Louie

Department of Physics, University of California at Berkeley, and Materials Sciences Division, Lawrence Berkeley Laboratory, Berkeley, CA 94720, USA

Abstract. A method for calculating the electronic and structural properties of solids using correlated wavefunctions together with quantum Monte Carlo techniques is described. The approach retains the exact Coulomb interaction between the electrons and employs a many-electron wavefunction of the Jastrow-Slater form. Several examples are given to illustrate the utility of the method. Topics discussed include the cohesive properties of bulk semiconductors, the magnetic-field-induced Wigner crystal in two dimensions, and the magnetic structure of bcc hydrogen. Landau level mixing is shown to be important in determining the transition between the fractional quantum Hall liquid and the Wigner crystal. Information on electron correlations such as the pair correlation functions which are not accessible to one-electron theories is also obtained.

1. Introduction

A major issue in calculating accurately the properties of solids from first principles is the treatment of electron-electron interactions or electron correlation effects. With very few exceptions [1-3], modern *ab initio* electronic structure calculations for solids treat this problem using some sort of self-consistent field schemes such as the Hartree-Fock (HF) or the local density functional (LDA) methods. These methods have been successfully applied to study the properties of many materials, but they have also encountered difficulties because of the approximate exchange-correlation functions used. In this paper, we describe some recent advances in using correlated many-electron wavefunction together with quantum Monte Carlo techniques to calculate materials properties. This approach allows treatment of electron-electron interactions in condensed matter systems going beyond self-consistent field schemes.

The present approach [1] employs a wavefunction of the Jastrow-Slater form and the exact Coulomb interaction between valence electrons. We discuss several applications briefly in this short review. For bulk semiconductors, calculations have been carried out for the electronic and structural properties yielding results in excellent agreement with experiment. The method has further been applied to compute quantities such as the electron pair correlation functions and single-particle orbital occupancy which are not accessible in standard self-consistent-field theories.

For the two-dimensional (2D) electron system, we applied the approach to investigate the phenomenon of Wigner electron crystallization. Recent experimental evidence [4,5] indicate that this elusive crystal is perhaps finally observed in 2D electron and hole systems at semiconductor heterojunctions in the fractional quantum Hall regime. The ground-state energy of a triangular Wigner crystal has been calculated as a function of carrier density, carrier mass, and applied magnetic field. The effects of exchange, correlations, and Landau level (LL) mixing are investigated. Our results show that LL mixing effects significantly alter the critical LL filling factor ν_c for the liquid-solid transition in the regime of experimental parameters.

Springer Series in Solid-State Sciences, Volume 114
Interatomic Potential and Structural Stability Eds.: K. Terakura and H. Akai

As a final example, we discuss the magnetic structure and equation of state of bcc solid hydrogen. This is studied as a prototypical Mott system with the long-range Coulomb interaction.

2. Correlated Wavefunction Variational Quantum Monte Carlo Approach

The use of correlated wavefunctions for many-electron systems dated back to the early days of quantum mechanics. The basic idea is to evaluate the ground-state properties of a system using the exact Hamiltonian but with a variational many-electron wavefunction. In the past, for condensed matter systems, various additional approximations had to be introduced to make the evaluation of the wavefunction and various physical quantities tractable. With the advance of Monte Carlo integration algorithms and modern computers, it is now possible to perform calculations on various systems without these further approximations.

In the present studies, the Hamiltonian is of the form

$$H = \sum_{i=1}^{N} \left\{ \frac{\hbar^2}{2m} \nabla_i^2 + V_{ext}(\mathbf{r}_i) + \frac{1}{2} \sum_{j \neq i} \frac{e^2}{r_{ij}} \right\} \tag{1}$$

where the last term is the exact Coulomb interaction between the electrons. For valence electron properties, the second term which describes the interaction with the cores may be accurately approximated by norm-conserving ionic pseudopotentials [1]. The many-electron wavefunction employed is of the form of an exponential correlation factor, the Jastrow factor, multiplying a Slater determinant of single particle orbitals:

$$\psi(\mathbf{r}_1,\ldots,\mathbf{r}_N) = \exp\left\{ \sum_{i=1}^{N} \chi(\mathbf{r}_i) - \sum_{i<j} u(r_{ij}) \right\} D(\mathbf{r}_1,\ldots,\mathbf{r}_N) \tag{2}$$

where N is the number of electrons in the system. A single particle term χ and a two-particle term u are retained in the Jastrow factor. The two-particle term correlates the motion of the particles so that two electrons may avoid each other dynamically, and the one-particle term allows a relaxation of the charge density distribution in the presence of the multiple-particle correlations. In the calculations, variational parameters enter into χ and u. The exact form of these correlation terms will depend on the system under consideration. However, physical arguments may often be used to contrain their general behaviors [1]. For accurate solid-state properties, simulation cell containing hundreds of electrons with periodic boundary conditions is required. Thus, it is necessary to use Metropolis Monte Carlo integration algorithms [6] in evaluating various physical quantities.

The correlated wavefunction VQMC method has shown to be a viable approach to the studies of electron correlation effects in solids. Calculations have been carried out for the cohesive and structural properties and correlation functions in various systems [1,3,7,8]. A particularly interesting and important recent application of this approach, which is not easily feasible in other *ab initio* schemes, has been the study of the effects of high magnetic fields [9].

3. Cohesive Properties and Electron Correlations in Solids

A first application of the method described in Sec. 2 to crystals was to the covalent solids. The binding energy and structural properties of carbon- and silicon-based crystals [1] were

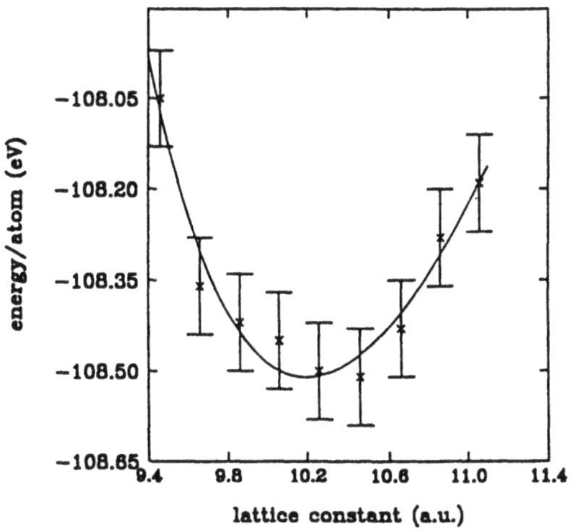

Fig. 1. Calculated energy of silicon (dots) together with a fit to the Murnaghan equation of state (line).

Table I. Calculated cohesive energies (in eV).

	LDA	VQMC	Experiment[a]
Diamond	8.63	7.45(7)	7.37
Graphite	8.65	7.40(7)	7.39
Silicon	5.29	4.81	4.7±0.1

a) Ref. 10

calculated using simulation cells with periodic boundary conditions containing up to 216 electrons (or 54 atoms). The Slater determinant part of the wavefunction was formed with single-particle orbitals obtained in a LDA calculation with *ab initio* norm-conserving pseudopotentials. A two-particle term of the form $u(r_{ij}) = A[1 - \exp(-r_{ij}/F)]/r_{ij}$ was used with A and F spin-dependent variational parameters. A single-particle term with one variational parameter which permits the valence charge density distribution to relax was also included.

Figure 1 shows the calculated total energy of silicon as a function of lattice constant. The theoretical results are also fitted to a Murnaghan equation of state. A calculated equilibrium lattice constant of 5.40 ± 0.04 Å and a bulk modulus of 108 ± 10 GPa are obtained, in very good agreement with the experimental values of 5.43 Å and 99 GPa, respectively. Similarly accurate results for the structural parameters have been obtained for diamond. The calculated cohesive energies of diamond, graphite, and silicon are presented in Table I. The correlated wavefunction results are in excellent agreement with experiment [10]. In general, Hartree-Fock (HF) calculations significantly underestimate the cohesive energy whereas the LDA calculations tend to overestimate the cohesive energy, typically by 15 - 20% or more. Electron correlation effects, thus, play a very significant role in determining the crystal cohesive energy. For example, the valence electron correlation energy is calculated to be 4.1 eV per atom in diamond but only 2.4 eV

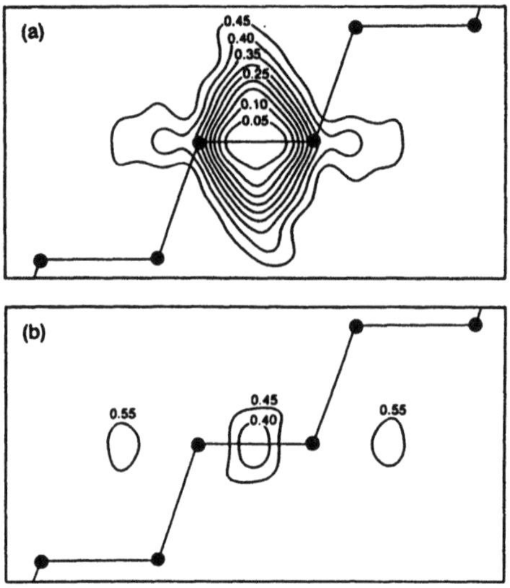

Fig. 2. Pair-correlation function in diamond for (a) parallel spin and (b) opposite spin with one electron at the bond center.

for the isolated carbon atom. In general, the correlated wavefunction VQMC calculations yield 90 - 95% of the correlation energy as compared to experiment.

With the wavefunction given in Eq. (2), it is straightforward to compute during the Monte Carlo sampling various quantities such as the single-particle orbital occupation number, the quasiparticle excitation energies within a Feynman single mode approximation, and the electron pair correlation functions [7]. For example, the pair correlation function $g_{\alpha\beta}(\mathbf{r}_1, \mathbf{r}_2)$ may be evaluate through the relation

$$g(x_1,x_2)n(x_1)n(x_2) = N(N-1)\int |\psi(r_1 = x_1, r_2 = x_2, \ldots, r_N)|^2 dr_3 \ldots dr_N \tag{3}$$

where n is the electron number density and spin indices are suppressed. Figure 2 illustrates the calculated pair correlation functions in diamond. Our results for $g_{\alpha\beta}$ of diamond and silicon show that, unlike the uniform electron gas case, g is indeed highly anisotropic and is a sensitive two-point function of $\mathbf{r}_1$ and $\mathbf{r}_2$. Both $g_{\uparrow\uparrow}$ and $g_{\uparrow\downarrow}$ are very rich in structure. For example, as seen in Fig. 2, the correlation hole, $h_{\uparrow\downarrow}(\mathbf{r}) = g_{\uparrow\downarrow}(\mathbf{r}_1,\mathbf{r}) - 1/2$, for $\mathbf{r}_1$ located at the bond center of diamond has a density distribution which is distinctly related to the structure and covalent character of the material. It is negative near $\mathbf{r}_1$, the bond center, as expected, but $h_{\uparrow\downarrow}$ is positive only in the nearby low density antibonding/interstitial regions and not in the neighboring bond centers reflecting the covalent nature of diamond. Since the pair correlation function is intimately related to the exchange-correlation energy density, information such as those contained in Fig. 2 may be used to gain microscopic understanding of electron correlations in solids.

The occupation number of a single-particle orbital ϕ_i is the ith diagonal element of the matrix

$$n(i,j) = N\int \phi_i^* (\mathbf{r}_1)\psi(\mathbf{r}_1,\mathbf{r}_2,\ldots)\psi^*(\mathbf{r}_1',\mathbf{r}_2,\ldots)\phi_j(\mathbf{r}_1')d\mathbf{r}_1 d\mathbf{r}_1' d\mathbf{r}_2 \ldots \quad . \tag{4}$$

The matrix n(i, j) again may be calculated by Monte Carlo evaluation. The quasihole energies within the Feynman single-mode approximation [11] are given by $\varepsilon_i = \langle\psi|c_i^\dagger[H,c_i$

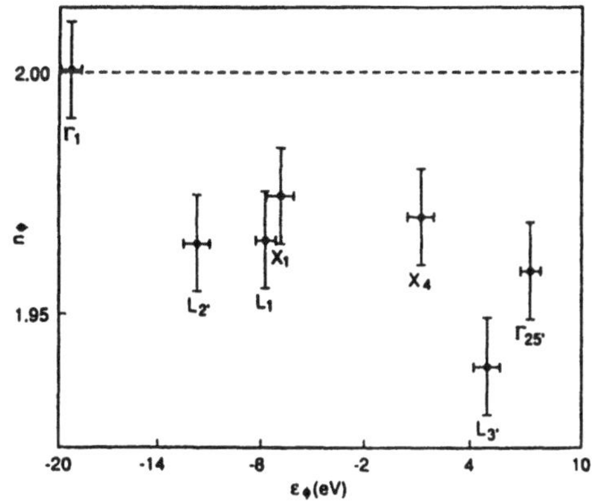

Fig. 3. Occupation of states in diamond below the Fermi level vs the single-mode-approximation energy of the state.

$]|\psi\rangle/n(i, i)$, where $c_i^\dagger$ and c_i are the single-particle creation and annihilation operators for the state ϕ_i, and $[H, c_i]$ is the commutator of the many-body Hamiltonian with the annihilation operator. This approximation depends for its success on the renormalization of the quasiparticle peak in the spectral distribution being negligible.

Some calculated single-particle orbital occupation numbers of the LDA orbitals (which have been shown to be virtually identical to the quasiparticle wavefunction [13]) versus corresponding quasihole energy for diamond are shown in Fig. 3. Complete occupation of a state would give an occupation number of 2. The results show the expected trend of lower occupation as one goes nearer the Fermi level, but the trend is not monotonic. The theoretical bandwidth is 25.4 ± 0.8 eV, in good agreement with the x-ray photoemission spectroscopy measurement [12] of 24.2 ± 1 eV but somewhat larger than other quasiparticle calculations [13].

4. Wigner Crystallization in the Fractional Quantum Hall Regime

In this section, we discuss the effects of exchange, correlations, and Landau level mixing to the ground-state energy of a triangular electron crystal in a perpendicular magnetic field [9]. In 1934, Eugene Wigner pointed out that, in the jellium model, electrons will crystallize at sufficiently low densities owing to the Coulomb repulsion which tends to localize the electrons. This quantum phase transition as a function of density at $T = 0$ has yet to be observed in three dimensions. However, there are several recent experiments [4,5] indicating that in the fractional quantum Hall regime, because of the strong localizing effect of the external applied magnetic field, Wigner crystallization is perhaps finally observed in 2D electron and hole systems at semiconductor heterojunctions.

The 2D electron or hole gas in a strong magnetic field, realizable experimentally in a MOSFET or a semiconductor heterojunction, exhibits a rich variety of phenomena [14]. By varying the carrier density, the carrier effective mass, and the strength of the external magnetic field, a very intricate phase diagram is expected. Among the various possible phases are the incompressible quantum Hall liquid phase and the Wigner crystal phase. Although the experimental evidence for the Wigner crystal phase are still being scrutinized, the observed re-entrant behavior to an insulating phase near Landau level filling factor $\nu = 1/5$ for n-type samples [4] and near $\nu = 1/3$ for p-type samples [5] of GaAs/AlGaAs junctions of comparable carrier densities have been interpreted as evidence for pinned Wigner crystals.

Exchange-correlation effects in 2D systems are fundamentally affected by the presence of a strong perpendicular magnetic field. For example, at fractional filling factors, particles can and prefer to correlate with each other to lower the interaction energy at no cost to the kinetic energy owing to the high degeneracy of the Landau levels. The presence of a strong B field introduces a new length, the magnetic length given by the radius of the Larmor orbit $\ell_B^2 = \hbar c/eB$, and a new energy, the cyclotron frequency $\omega_c = eB/m^*c$, into the problem. Both quantities are independent of the carrier density. Crystallization is expected when ℓ_B is less than the inter-electron distance d, that is, in the regime of $\upsilon < 1$. Landau level mixing is expected to be important when the electron-electron interaction energy $E_{coul}=e^2/\varepsilon d$ is comparable to or larger than the inter-LL spacing $\hbar\omega_c$. There are, therefore, three important materials parameters in determining the ground state: the density n, the effective mass m*, and the applied B field. The system is characterized by two independent dimensionless quantities which are the filling factor υ which is a function of n and B and the electron gas parameter r_s which is a function of n and m*. There have been many studies on B-field induced Wigner crystals in 2D [14-18]. Most of these were carried out, however, within the Hartree-Fock approximation. With few exceptions [18], only the lowest Landau level is considered.

In our simulation, the usual periodic boundary conditions cannot be directly applied to electrons in a strong applied B field due to the aperiodic vector potential in the Hamiltonian. For example, in the symmetric gauge, $\mathbf{A} = (-yB/2, xB/2)$ for a uniform field. The appropriate translational symmetry may be recovered by introducing the concept of rational fields [19] and noting that the physics of the system should be gauge invariant and that a finite translation in the present case is equivalent to a gauge transformation. These symmetry properties allow us to perform simulation with modified periodic boundary conditions in a magnetic field.

Our trial wavefunctions for the Wigner crystal are still in the general Jastrow-Slater form of an exponential correlation factor multiplying a function of the single-particle orbitals. The single-particle orbitals are chosen to be Gaussians localized about the lattice sites:

$$\phi_j(\mathbf{r}) = \frac{1}{\sqrt{2\pi}} \frac{\beta}{\ell_B} \theta(\mathbf{R}_j) \sum_{\{\mathbf{T}\}} \theta(\mathbf{T}) \exp\left\{-\frac{\beta^2 r^2}{4\ell_B^2}\right\} . \qquad (5)$$

Here β is a variational parameter involving LL mixing and to be optimized. For $\beta = 1$, Eq. (5) corresponds to a wavefunction in the lowest LL localized on a lattice site $\mathbf{R}_j$ with the proper phase factors resulting from the use of a symmetric gauge at the origin. (The **T**'s are translation vectors of the simulation cells associated with the periodic boundary conditions used and θ is a magnetic translation operator.) All calculations reported are done for spin-aligned electrons with a 100-electron simulation cell. Tests show that the resulting finite size effects are smaller than the statistical noise in the results.

We first summarize the results from wavefunctions which are made up of functions only from the lowest Landau level with either exchange or correlations included and then discuss next the effects of LL mixing. This first part of the discussion serves to make contact with earlier work and to establish the validity of the correlated wavefunction VQMC method. Our exchange-only (or HF) results are obtained by using a wavefunction which is just a Slater determinant of the single-particle orbitals given above with $\beta = 1$. The correlation-only results are calculated for distinguishable particles (again, with $\beta = 1$ for the orbitals) with the Jastrow correlation factor derived from the consideration of the magnetophonons similar to the work of Lam and Girvin [16]. In the large field limit, this Jastrow factor may be written as an analytic function of the complex variables which describe the position of the particles in 2D. Hence, in both calculations, there is no mixing in of higher LL's into the ground-state wavefunction by construction, and the kinetic energy per electron is trivially given by $\hbar\omega_c/2$.

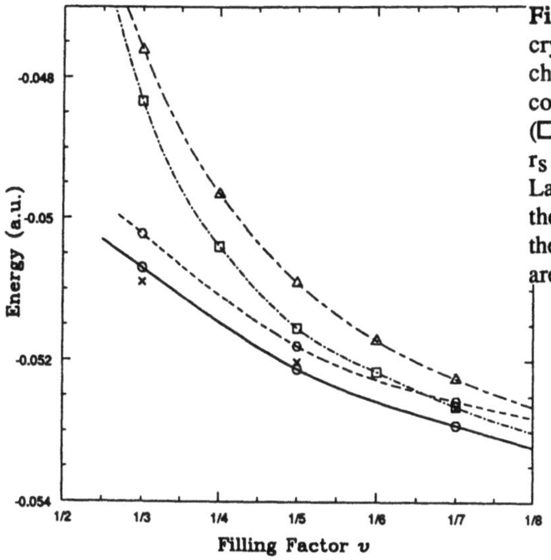

Fig. 4. Calculated Wigner electron crystal energy E - $\hbar\omega_c/2$ using exchange only with no LL mixing (Δ), correlations only with no LL mixing (□), correlations with LL mixing for $r_s = 20$ (—). The energy of the Laughlin liquid (Ref. 17) is given by the dashed curve, and the energies for the liquid with LL mixing (Ref. 21) are given by the crosses.

The exchange-only and correlation-only results with no LL mixing are presented in Fig. 4 by the curves with the triangle and square symbols, respectively. Previously, the most accurate energies for the 2D electron system in a strong magnetic field in the exchange-only approximation were calculated from the viewpoint of charge density waves (CDW) [15] rather than the explicit Wigner crystal wavefunction used here. Our present HF results are extremely close to the CDW results. Thus, little is gained by the explicit self-consistency procedure used in the CDW approach.

Our correlation-only results with no LL mixing are the same as those obtained by Lam and Girvin [16] who used a special k-point sampling scheme in calculating the energy rather than the VQMC method. In this approximation, the Wigner crystal is considered as distinguishable particles correlated by the magnetophonons. Thus, although it contains correlation effects, exchange interaction is ignored. As noted in Ref. 16, at $\upsilon = 1/2$, the correlation-only result is, in fact, higher in energy than the HF result due to the rather large exchange interaction at this filling factor. However, a comparison of the Hartree, the exchange-only, and the correlation-only energies in the lowest LL approximation shows that correlation effects dominate over exchange in the range of $\upsilon \lesssim 1/3$. The energies for the Laughlin state taken from Ref. 17 are given by the dashed curve with the hexagon symbols in Fig. 4. Thus, within the lowest LL approximation, the cross-over from the liquid state to the solid state is $\upsilon_c \approx 1/6.5$ as found by Lam and Girvin.

Now let us discuss the effects of Landau level mixing. Our work is motivated by the recent experimental work on the 2D hole systems [5] which indicates the occurrence of a Wigner crystal phase around $\upsilon = 1/3$. This earlier transition from that of the electron case is ascribed to a larger LL mixing due to the heavier hole mass in GaAs. Even for the electron systems, the electron interaction energy is comparable to the cyclotron frequency. A relevant measure of the importance of LL mixing is the ratio $E_{coul}/\hbar\omega_c = \upsilon r_s/2$. A higher value for the ratio indicates a strong LL mixing. For typical 2D electron systems $r_s \sim 2$, but for the 2D hole system investigated, $r_s \sim 13$ - 25 depending on the assumed hole effective mass [5,20].

Admixture of higher LL orbitals to the ground state allows a lowering of the interaction energy at the expense of the kinetic energy. Because of the Coulomb repulsion between electrons on neighboring sites, it is energetically favorable to have (1) a charge density distribution more localized than the one given by single-particle orbitals in the lowest Landau level and (2) a nonanalytic correlation term u in the Jastrow factor which optimizes the dynamical short-range avoidance of two electrons. These are the driving mechanisms for Landau level mixing. We put in these two effects by optimizing the energy with respect to the parameter β in Eq. (5) and by adding to the magnetophonon correlation factor a term of the form $u(r) = A[1- \exp(- \sqrt{r/F} - 1/2\ r/F)]/\sqrt{r}$ where r is the distance between two electrons. Both the deviation of β from 1 in the single-particle orbitals and the additional term in the Jastrow exponent correspond to admixture of higher LL's into the wavefunction, and one needs to evaluate both the kinetic and interaction energies. The changes in energy and in the localization of the electrons due to LL mixing can be quite large in the range of $\upsilon = 1/2 - 1/5$ depending on the value of r_s. For example, at $\upsilon = 1/3$, with just an optimization in the parameter β, we find an increase in density at the lattice site by $\delta\rho(0)/\rho(0) = 70\%$ and an lowering in energy by $\delta E/(E- \hbar\omega_c/2) = - 4.4\%$ at $r_s = 20$ whereas at $r_s = 2$ the same two quantities are 10% and - 0.8%, respectively. With the LL-mixing Jastrow factor, the total energy is lowered even further.

The final results with full LL mixing and magnetophonon correlations for the case of $r_s = 20$ are given in Fig. 4 as the solid curve. The parameters used in the calculation correspond roughly to those in the experiment on the 2D hole systems [5]. We see that there is a dramatic lowering in energy by allowing LL mixing in the range of υ of experimental interest. At $r_s = 20$, LL mixing effects are, in fact, larger than those of intralevel exchange-correlation effects. And as seen in Fig. 4, the Wigner crystal energy curve lies entirely below the Laughlin wavefunction energies at $r_s = 20$ for the whole range of υ considered. This is not an accurate comparison of the energy of the two phases because the Laughlin wavefunction [17] does not include higher LL's. There is, however, a recent VQMC calculation [21] on the fractional quantum Hall liquid using a Laughlin-like wavefunction but included LL mixing. The calculations were carried out on a sphere with a LL-mixing correlation term in the Jastrow factor. Their results for the case of $r_s = 20$ are presented in Fig. 4 by the cross symbols. As expected, LL mixing does not affect the liquid phase as much as the solid phase. Comparison of the two calculations shows that the liquid-solid transition occurs near $\upsilon = 1/3$. Thus, our results give quantitative support to the observation that, because of LL mixing effects, 2D hole systems at GaAs/AlGaAs interface crystallize at a larger υ than 2D electron systems at comparable densities.

5. Magnetic Structure of BCC Hydrogen

The present approach has also been applied to study the equation of state and magnetic structure of bcc solid hydrogen [8,22]. Various forms of the wavefunction have been used to investigate the relative stability of the paramagnetic (PM) state, the antiferromagnetic (AFM) state, and the ferromagnetic (FM) state as a function of the lattice constant. The calculations were carried out at densities between Wigner-Seitz radius of $r_s = 1.0$ to 3.0 a.u.

The results of the 54-atom simulation cell calculations [8] are presented in Fig. 5. Most of the interesting features occur between $r_s = 2.0$ and $r_s = 3.0$. An earlier LDA study [23] suggested that both the magnetic and metal-insulator transitions occur in this region. Our results show that the fully FM phase is unbound at all densities which is analogous to the unbound hydrogen molecule in a spin-triplet electronic state. Among the magnetic structures considered, the PM phase is the ground state for r_s less than 2.2, and its energy rises above the fully FM phase at $r_s = 2.7$. Contrary to some previous work [23,24], we did not find any indication of a stable partial-FM phase at the densities considered.

Our results show a transition from PM to AFM phase at $r_s = 2.2$. At this density the system also undergoes a metal-insulator transition with an energy gap opening up in the

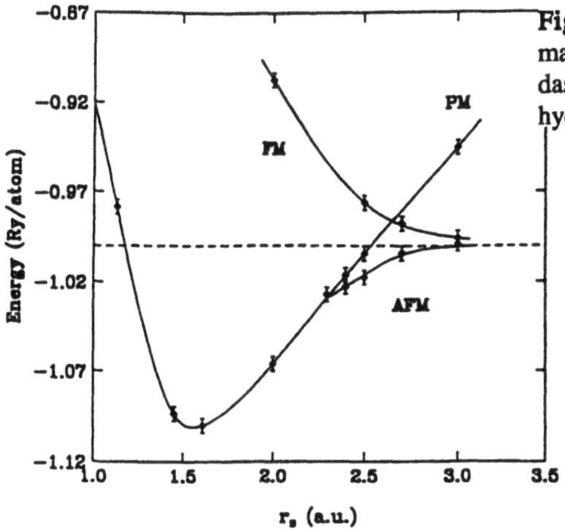

Fig. 5. Energy vs. density for three magnetic structures of bcc hydrogen. The dashed line is the energy of an isolated hydrogen atom.

excitation spectrum. The present transition density of $r_s = 2.2$ is significantly different from previous LDA [23] or self-interaction corrected LDA [24] calculations. Also, as for other materials, the cohesive energy is significantly overestimated in the LDA calculation. For example, at $r_s = 3.0$, the LDA calculations found that the AFM phase is bound by 15 mRy/atom while our VQMC results show that the binding is less than 1 mRy/atom.

6. Summary

We have described a correlated wavefunction variational quantum Monte Carlo method for calculating the total energy and related properties of solids. Highly accurate cohesive and structural properties, as well as quantities such as the electron pair correlation functions, have been obtained for carbon- and silicon-based crystals. The energetics of the magnetic-field-induced Wigner crystal in 2D in the fractional quantum Hall regime has also been investigated. Landau level mixing effects are shown to be very important in the range of density and magnetic field strength of experimental interest. The results provide an explanation for the recent observation of a re-entrant insulating transition near $\upsilon = 1/3$ for 2D hole systems at GaAs/AlGaAs interfaces. We also investigated the magnetic structure and metal-insulator transition in bcc hydrogen as a function of density. The present approach is shown to be quite versatile and provides a viable new avenue for computing the properties of and studying the electron-electron interactions in a variety of solid-state systems.

Acknowledgments: This work was supported by NSF Grant No. DMR91-20269 and by the Director, Office of Energy Research, Office of Basic Energy Sciences, Materials Science Division of DOE under Contract No DE-AC03-76SF00098. SGL acknowledges collaboration with S. Fahy, X. W. Wang, J. Zhu, and X. J. Zhu on the work reviewed in this paper. Supercomputer time was provided by NSF at the San Diego Supercomputer Center and by DOE at the Livermore NERSC.

References

[1] S. Fahy, X. W. Wang, and S. G. Louie, *Phys. Rev. Lett.* **61**, 1631 (1988); *Phys. Rev. B* **42**, 3503 (1990); **43**, 9299 (E) (1991).

[2] B. Kiel, G. Stollhoff, C. Weigel, P. Fulde, and H. Stoll, *Z. Phys. B* **46**, 1 (1982); G. Stollhoff and K. P. Bohnen, *Phys. Rev. B* **37**, 4678 (1988).

[3] X. P. Li, D. M. Ceperley, and R. M. Martin, *Phys. Rev. B* **44**, 10929 (1991).

[4] E. Y. Andred *et al., Phys. Rev. Lett.* **60**, 2765 (1988); H. W. Jiang *et al., ibid.* **65**, 633 (1990); V. J. Goldman *et al., ibid.* **65**, 2189 (1990); Buhmann *et al., ibid.* **66**, 926 (1991); H. W. Jiang *et al., Phys. Rev. B* **44**, 8107 (1991).

[5] M. B. Santos *et al., Phys. Rev. Lett.* **68**, 1188 (1992).

[6] N. Metropolis, A. W. Rosenbluth, M. N. Rosenbluth, A. H. Teller, and E. Teller, *J. Chem. Phys.* **21**, 1087 (1953).

[7] S. Fahy. X.-W. Wang, and S. G. Louie, *Phys. Rev. Lett.* **65**, 1478 (1990).

[8] X. W. Wang, J. Zhu, S. G. Louie, and S. Fahy, *Phys. Rev. Lett.* **65**, 2414 (1990).

[9] X. J. Zhu and S. G. Louie, to be published.

[10] There is a large scatter in the experimentally measured cohesive energy of Si. See, for example, Ref. 1 and references therein.

[11] R. P. Feynmann, *Phys. Rev.* **94**, 262 (1954).

[12] *Landolt-Bornstein: Numerical Data and Functional Relationships in Science and Technology*, ed. O. Madelung (Springer-Verlag, New York, 1982), Vol. 17.

[13] M. S. Hybertsen and S. G. Louie, *Phys. Rev.* **B 32**, 7005 (1985); S. Horsch, P. Horsch, and P. Fulde, *ibid.* 1870 (1984);

[14] *The Quantum Hall Effect*, eds. R. E. Prange and S. M. Girvin (Springer-Verlag, New York, 1990), 2nd ed.

[15] D. Yoshioka and H. Fukuyama, *J. Phys. Soc. Jap.* **47**, 394 (1979); D. Yoshioka and P. A. Lee, *Phys. Rev. B* **27**, 4986 (1983); K. Maki and X. Zotos, *Phys. Rev. B* **28**, 4349 (1983).

[16] P. K. Lam and S. M. Girvin, *Phys. Rev. B* **30**, 473 (1984); **31**, 613 (E) (1985).

[17] R. B. Laughlin, *Phys. Rev. Lett.* **50**, 1395 (1983); D. Levesque, J. J. Weis, and A. H. MacDonald, *Phys. Rev. B* **30**, 1056 (1984).

[18] A. H. MacDonald, *Phys. Rev. B* **30**, 4392 (1984); D. Yoshioka, *J. Phys. Soc. Jpn.* **53**, 3740 (1984); **55**, 885 (1986); K. Esfarjani and S. T. Chui, *J. Phys. C* **3**, 5825 (1991)

[19] See, for example, J. Callaway, *Quantum Theory of the Solid State* (Academic Press, New York, 1974).

[20] H. L. Stormer *et al., Phys. Rev. Lett.* **51**, 126 (1983).

[21] R. Price, P. M. Platzman, and S. He, to be published.

[22] J. Zhu, X. W. Wang, and S. G. Louie, to be published.

[23] B. I. Min, T. Oguchi, H. J. F. Jansen, and A. J. Freeman, *Phys. Rev. B* **33**, 324 (1986).

[24] A. Svane and O. Günnarson, *Solid State Comm.* **76**, 851 (1990)

Quantum Monte-Carlo Calculations on Real Materials

R.M. Martin[1,2], *Y. Kwon*[1,2], *X.-P. Li*[1,2;*], *L. Mitáš*[1,3], *V. Natoli*[1,2], *E.L. Shirley*[1,2;**], *and D.M. Ceperley*[1,3]

[1]Physics Department, University of Illinois at Urbana-Champaign, 1110 W. Green Street, Urbana, IL 61801, USA

[2]Material Research Laboratory, University of Illinois at Urbana-Champaign, 1110 W. Green Street, Urbana, IL 61801, USA

[3]National Center for Supercomputer Applications, University of Illinois at Urbana-Champaign, 1110 W. Green Street, Urbana, IL 61801, USA

*Present address: Department of Physics, Rutgers University, Piscataway, NJ 08855, USA

**Present address: Department of Physics, University of California, Berkeley, CA 94720, USA

Quantum Monte Carlo methods can calculate exact or nearly exact ground state properties of many body systems. The subject of this paper is a brief outline of some of the current developments in applying such methods to general materials and results of recent calculations. A critical issue which we discuss is the accurate treatment of core electrons, which is crucial to *ab initio* calculations on materials.

1. INTRODUCTION

One of the major goals of current theoretical work is the development of computational methods for accurate *ab initio* calculations of interesting properties of "real" materials, such as, molecules, crystals, surfaces, defects, artificial structures, and other types of materials of scientific and technological relevance. Since the properties are determined by the many-body system of electrons interacting with the nuclei, truly *ab initio* methods must deal with this many-body problem. Quantum Monte Carlo[1] (QMC) has an important role in reaching this goal because the statistical sampling methods used in Monte Carlo methods can simulate exactly (or nearly exactly) large many-body systems. Configuration interaction methods can find exact ground and excited states; however, they require computational time which scales as $N!$ where N is the number of particles. In contrast, Monte Carlo simulations of the ground state of a many-body system scale as N^3 so that they can be applied to large enough systems that they can be extrapolated to the bulk limit. In this paper we will consider QMC as a method to determine only ground state properties (even though certain excited states can also be treated by QMC and projection techniques are beginning to be developed for excitations[2]). Nevertheless, this is a very important goal at the present time because the electronic

Interatomic Potential and Structural Stability Eds.: K. Terakura and H. Akai

ground state determines the atomic structure, phase transitions, magnetic order at T=0, elastic properties, defect energies, phonon frequencies, and a host of other properties. In addition to the intrinsic interest of such ground state properties, determination of the atomic scale structure is a crucial step for *ab initio* calculations of excited electronic states.

The work described here is part of our efforts to make it possible to carry out such nearly exact calculations on general condensed matter systems. We will concentrate on variational and diffusion QMC methods, both of which have crucial roles in the electronic structure problem. Since it is already well established that such systems as hydrogen and the homogeneous electron gas can be treated by QMC, the primary new developments needed are: efficient representations of the many-body trial wavefunctions for electrons in the highly inhomogeneous environments of atoms, surfaces, etc. and accurate ways to treat the core electrons of elements heavier than He. These are the major subjects discussed here.

2. QUANTUM MONTE CARLO METHODS AND THE CORE PROBLEM

Among the many different types of QMC methods,[1] we will focus upon Variational Monte Carlo[3, 4] (VMC) and Green's Function[5, 6] (or Diffusion[7]) Monte Carlo (DMC), which are capable of treating real electron problems with the accuracy needed in condensed matter problems. VMC uses statistical sampling methods to evaluate expectation values of any operator with a correlated many-body trial wavefunction; by optimizing parameters in the wavefunction, one can find an upper bound to the exact energy and determine which types of correlations are most important among the electrons. The two great advantages of VMC are: 1) its computational and conceptual simplicity, and 2) the explicit analytic representation for the wavefunction, which manifestly includes some correlations but not others. The primary disadvantages are: 1) the desired correlations must be put in, and 2) there is no intrinsic way to check the assumptions. Since the energy is variational one can often find good estimates of the energy; conversely, it is difficult to use energy minimization alone to find the wavefunctions since the energy is insensitive to the errors in the wavefunction.

DMC starts from a variational function and iterates the Schrodinger equation in imaginary time to project out the best possible ground state energy and correlation functions.[1, 5, 6] For Boson systems, because the ground state has no nodes, QMC methods can find the exact ground state energy and all other ground state properties. For Fermions, however, the famous "sign problem" [1] - due to the antisymmetry of the particles under exchange - has prevented the formulation of an exact method that is feasible for more than a few particles.[8] Nevertheless, the "fixed node approximation" [1] is extremely accurate if one has a trial function with appropriate nodes. The accuracy has been established by carrying out "release node" calculations for many systems, including the homogeneous electron gas, small molecules, and solid hydrogen. [5, 6, 9] The primary advantages of DMC are: 1) its ability to find exact (or nearly exact) ground state energies and correlations beyond those included in the trial wavefunction, and 2) a greatly decreased sensitivity to the choice of the trial function. Its primary disadvantages are: 1) the wavefunction is represented only implicitly in the sampling and 2) difficulties in using the fixed-node method in conjunction with non-local pseudopotentials which arise in treating atoms with cores, as will be discussed below.

Although core electrons are relatively inert, their presence has crucial effects upon the active valence electrons. The core states are problematic in sampling methods because their characteristic energy scales are large and time scales small; in DMC this causes an increase in computational time to achieve a given accuracy for the total energy which scales as $Z^{6.5}$, where Z is the atomic number.[10] The reasons for this rapid increase with Z is that the time scales as the cube of total number of electrons Z^3, with additional factors of Z^2 because the step size in the simulation must be decreased and $Z^{1.5}$ because of the increase in total energy. For this reason, direct calculations are not feasible even for heavy atoms, let alone molecules or solids! Our approach involves full many-body calculations on the valence electrons only, with effects of the core electrons replaced by a pseudopotential[11] (PP) or pseudohamiltonian[10, 12] (PH). This breaks the problem into two parts: a) finding accurate pseudopotentials and b) simulating valence electrons moving in the non-local potentials which are more complex than ordinary local pseudopotentials. Fahy, Wang, and Louie[4] have carried out non-local pseudopotential calculations using VMC methods, with remarkable agreement with experiment for the cohesive energy of carbon and silicon. Li, et.al.,[13] have carried out DMC calculations with a PH and shown that it is indeed possible to calculate the total valence energy to a precision of order 0.05eV per atom. This is sufficient for many real problems and leads to properties of Si in good agreement with experiment, including improvement of the well-known errors in the cohesive energy found in the local density approximation (LDA).

The second part of our work addresses the question: how can one generate a pseudopotential or pseudohamiltonian that treats the effects of the cores with sufficient accuracy that they are worthy to be used in such accurate valence calculations. Progress in this direction has been made in recent work,[14, 15] which has shown how to estimate the non-local term from a variational function while treating the local terms in the full fixed-node DMC. We also must face the fact that up to now pseudopotentials have been generated only in approximate one-electron methods such as LDA or Hartree Fock.[11] Any errors made in generating the potentials propagate directly into the final answers. To overcome this defect, we have recently developed methods for a more rigorous many-body "core-valence partitioning" that can be used to generate improved "quasiparticle pseudopotentials" [18, 19] which are described in Section 4.

3. CALCULATIONS ON SOLIDS

In VMC one minimizes the energy with respect to a trial function, which is often chosen to have the Jastrow-Slater form:

$$\Psi(\mathbf{r}_1,\ldots,\mathbf{r}_N) = \exp\left[\sum_{i=1}^{N}\chi(\mathbf{r}_i) - \sum_{i\leq j} u(r_{ij})\right] D(\mathbf{r}_1,\ldots,\mathbf{r}_N) \qquad (1)$$

Other more general correlated functions can be used and are essential to capture some important correlations.[15, 20] The VMC trial function is also the starting point for DMC[7] where the operator $\exp(-tH)$ projects out the ground state from the given trial function. Here $u(r)$ is a two body correlation function obtained from the random phase approximation for a homogeneous electron gas, $\chi(\mathbf{r})$ is a one body term which modifies the VMC charge density,[3, 4] and D is a Slater determinant of single particle states. In our work LDA calculations are used to construct good trial orbitals and to extrapolate to large cell size. In our work the orbitals are expanded in plan waves with energy below a cutoff.

The most extensive tests of the methods have been done on metallic hydrogen at high pressures.[16] There it was found that the Slater-Jastrow functions using LDA orbitals gives a significantly better variational energy than the simpler functions previously used by Ceperley and Alder.[9] (This was also found by Wang, et. al. [17] in their studies of H.) The DMC energy was lowered slightly due to improved nodes; however, the change is much less than for VMC. We have found that for low-coordination structures not considered by Ceperley and Alder,[9] it is important to use the LDA orbitals (and to occupy the states in the metallic system according to the LDA Fermi surface) in order to distinguish between the many structures with nearly the same energies. If the protons are held fixed then the VMC simulations for the electrons give energies above the DMC by a nearly constant amount ($\sim$ 0.004 Ry), i.e., VMC calculations with a Slater-Jastrow trial function predict accurately the ordering of the structures. However, if the protons are treated as quantum particles, then present VMC functions are not as accurate; the energy differences are larger and the variation ($\sim$ 0.005 to 0.012 Ry) means that the VMC results do not accurately distinguish the structures. In this case, the value of DMC is apparent.

As a first test on other solids, calculations have been done by two groups for silicon.[4, 13] Fahy, et.al.,[4] carried out VMC calculations using a non-local pseudopotential (PP), whereas Li, et.al.,[13] used a pseudohamiltonian[10] (PH) which can be employed directly in both VMC and DMC methods. We will discuss explicitly the work of Li, et. al.,[13] which was done with a cubic supercell containing 64 atoms with periodic boundary conditions. In the Slater-Jastrow wavefunction, two cutoffs for the LDA orbitals were used, 7 Ry and 15 Ry, to test the influence on the calculated energy. In VMC, the energy obtained with orbitals cutoff at 15 Ry is 0.44 eV lower than the energy with a 7 Ry cutoff, while the two DMC energies are the same within statistical errors (0.04 eV/atom.) This test suggests that the error in the nodal locations caused

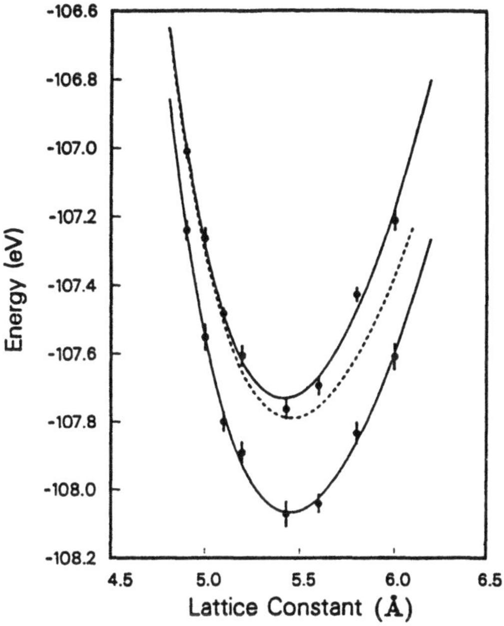

Figure 1: Total energy of silicon versus lattice constant using a pseudohamiltonian (PH) from VMC (upper curve), DMC (lower), and LDA (dashed). The same corrections for finite size effects have been applied to VMC and DMC points and error bars show the estimated statistical errors.

by the truncation of the LDA trial function is small. Results found using the larger cutoff are discussed here. The difference in an LDA calculation between a 64 atom and infinite system is 0.11 eV, and our QMC results are corrected assuming that they have the same size dependence as LDA. The number of walkers in the DMC ensemble is chosen to be 200 and the initial distribution was obtained from VMC. A time step of 0.015 in atomic units was used. (This gives an acceptance ratio of 98% in the Metropolis portion of the time evolution.) A test calculation using half the time step gave identical results, showing that the time step error is less than 0.03eV/atom. A typical run with 3×10^4 steps, took 20 hours of CRAY-XMP time. To achieve the same error bars, the DMC calculation takes only 2.6 as much computer time as VMC but does not require systematic search of trial functions.

Figure 1 shows the energy as a function of lattice constant from the LDA, VMC and DMC calculations.[13] The curves are least square fits to the Murnaghan equation of state. The total energy dropped 0.21(3) eV in the atom and 0.34(3) eV/atom in the solid at zero pressure (with the most accurate variational function) in going from the VMC to DMC. This difference reflects the fact that it is easier to construct a good trial function in the atom than in the solid. In VMC it is important to construct equally good trial functions at all the lattice constants, otherwise there will be a systematic bias in the results.

Table 1: Comparison of LDA, VMC, and DMC for crystalline silicon taken from Li, et.al.; [a] Using pseudopotential from Fahy, et. al.; [b] from Fahy, et. al. ; energies are in eV.

	a (Å)	B (Mb)	dB/dP	$E_{at.}$	$E_{sol.}$	E_{coh}
LDA PP[a]	5.373	0.946	4.00	-102.71	-108.01	5.30
LDA PH	5.455	0.916	3.93	-102.71	-107.79	5.08
VMC-PP[b]	5.40(4)	1.08(10)	—	-103.42(3)	-108.23(6)	4.81(7)
VMC-PH	5.42(2)	1.08(5)	3.5(6)	-103.35(3)	-107.73(2)	4.38(4)
DMC-PH	5.45(2)	1.03(7)	3.8(8)	-103.56(2)	-108.07(2)	4.51(3)
Exp.	5.430	0.992	3.2–4.7	—	—	4.63(8)

Also shown in Fig. 1 are the LDA total energies using the same pseudohamiltonian (PH). A comparison with experiment and other calculations is given in Table 1. For semiconductors like silicon, LDA is known to work very well, and indeed the total energies from LDA are very close ($\sim$ 0.2 eV) to those from QMC. A larger change occurs in the atom where the spin polarized LDA energy is ($\sim$ 0.8 eV) higher than DMC. The difference 0.6 eV is the correction to the LDA binding energy, bringing the computed result much closer to experment.

Our DMC cohesive energy, bearing in mind the unknown transferability of the PH, should be between 4.51 (assuming no correction) and 4.73 eV (assuming the LDA gives correctly the difference between the PH and PP), in general agreement with the most quoted experimental value, 4.63(8) eV. In future work, we plan DMC calculations with a non-local "quasiparticle" pseudopotential (described below) using a new DMC method,[14, 15] which treats the non-local parts in a variational manner. Together with evidence discussed below, this shows that accurate construction of the pseudopotentials from a many-body theory is necessary for the full accuracy of the QMC approach to be reached.

4. QUASIPARTICLE PSEUDOPOTENTIALS

The above discussion shows that the accuracy of QMC calculations applied to real solids is now limited by the quality of the pseudopotentials used to represent the effects of the cores. However, essentially all previous pseudopotentials have been derived using a one-electron method such as Hartree-Fock or LDA.[11] In one-electron methods, core and valence states are separate eigenfunctions and a pseudopotential can readily be derived which reproduces the valence function outside the core. In a many- body theory it is not clear *a priori* how to make such a separation since all electrons are identical. Shirley and coworkers have devised a way to do this "core/valence partitioning" taking into account the exchange and correlation between the core and valence electrons.[18]

The essence of this core/valence partitioning method is to treat the valence electrons as "quasiparticles" with self-energies which reflect the effects of the cores, i.e., exchange, dynamical correlation, and relaxation. These effects are calculated in the atom (or ion) using the Greens function method[21] which is a variant of Hedin's GW approximation, based upon the generalized RPA, which we term generalized GW (GGW). Diagrammatic summation is successful for describing the effects of cores because the large gap for core excitations makes the sums converge rapidly.[18] The result is that the Green's function for electron addition and removal in the range of valence energies can be described by valence electrons moving in the presence of frozen core orbitals plus a self-energy described by a "core polarization potential" V_{cp} which takes into account core relaxation and dynamic correlation. The form of V_{cp} was chosen following the work of Muller, et. al.[22] In addition, core polarization modifies the electron-electron interaction in a simple way near the ion cores. Finally, usual methods[11, 19] can be used to transform this to a non-local pseudopotential valence-only problem.

The first test of the "quasiparticle" pseudopotential is for one electron outside a core. In that case the valence problem is a simple one-electron problem and the eigenvalue of the "quasiparticle" pseudopotential should agree with the experimental binding energy of the electron in the exact many-body atom. Representative examples of results on atoms showing accuracy of our potential are given in Table 2. The row labelled GGW shows the results of full atom GGW calculation; the next row (GGW/PP) shows the small errors introduced in the pseudopotential transformation. Both agree very well with experiment compared to the single body methods listed below in the table. We see that the new quasiparticle pseudopotential has substantial improvements over LDA or HF derived potentials. The errors shown in Table 2 for the one electron potentials will propagate in any many-body valence calculation which uses them.

The goal, however, is to apply these potentials to many-valence-electron systems such as molecules and solids. As a test,[18] we have carried out a Monte Carlo simulation of the binding curve for the Na dimer - which gives the *exact* energy for this two-valence-electron system. Parallel calculations have

Table 2: Removal energies (in eV) for one electron bound to a core from experiment, the full atom generalized GW (GGW), the quasiparticle pseudopotential (GGW/PP) derived from the GGW, Koopman's eigenvalue (HF), self-consistent HF, and pseudopotential local spin density (LDA PP). (from Shirley, et. al.)

	Be $2s$	Be $2p$	Na $3s$	Na $3p$	Sc $4s$	Sc $4p$	Sc $3d$	Sc $4f$
Expt.	18.21	14.25	5.14	3.04	21.58	16.99	24.73	7.76
GGW	18.21	14.25	5.11	3.03	21.56	17.00	24.53	7.77
GGW/PP	18.21	14.25	5.12	3.03	21.58	17.01	24.54	7.77
HF	18.13	14.14	4.96	2.98	20.90	16.57	23.09	7.69
HF-SCF	18.13	14.14	4.96	2.98	20.97	16.60	23.76	7.69
LDF PP	18.30	14.61	5.30	3.20	21.34	16.91	25.12	- -

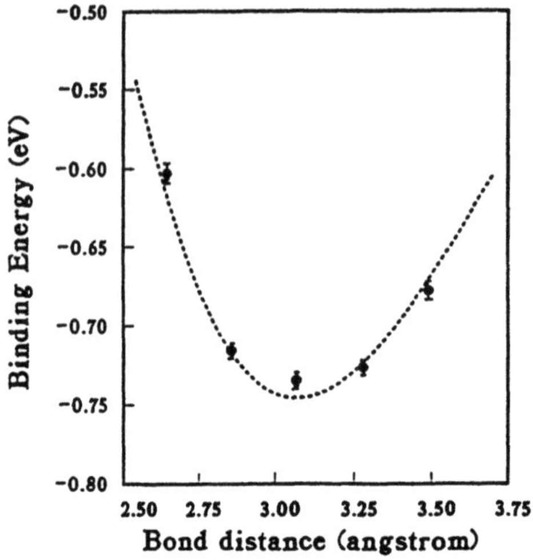

Figure 2: DMC energies with error bars for the Na_2 dimer using the GGW quasiparticle pseudopotential, compared to experiment. Comparison with HF and LDA pseudopotentials is described in the text

been done with HF, LDA, and GGW "quasiparticle" pseudopotentials. Figure 2 shows the nearly exact agreement of the QMC calculation using the GGW potential with the experimental binding curve for the dimer. In contrast, the curves with the HF generated pseudopotential are too weakly bound (0.02eV) with equilibrium distance about 5% too large. Results with an LDA generated potential[18] are overbound by a larger amount, 0.1 eV, and give too small a bond length.[19]

5. SURFACES

What has been done on materials, beyond perfect crystals and simple molecules? As a first step Li, et. al.[23] have carried out full VMC and DMC calculations on the simplest surface problem, the jellium surface. The case considered was jellium at the density of aluminum ($r_s = 2.1$), which was chosen because there are good arguments that this is an excellent approximation to real Al and because the results can be compared directly with other theoretical calculations. [24] For this density, the calculated surface energy from DMC is rather close to the LDA result (closer than to the previous many-body hypernetted chain calculations[24]) and is in remarkable agreement with recent gradient-corrected LDA calculations.[25] This is an example of the advantages of the DMC method. In this case, the variational calculations were rather poor because it is difficult to find good analytic representations for electron correlations near the surface. Although the VMC charge density is poor, this is corrected by DMC. Examples of the electron correlation and exchange holes and their distortion near the surface have been found;[23] these could serve as the basis for constructing new improved correlated wavefunctions for electrons at the surfaces of metals.

6. CONCLUSIONS

There are two primary conclusions of our work. The first is that given a pseudopotential or pseudohamiltonian, DMC calculations for valence electrons only can be carried out with an absolute accuracy of order 0.05 eV per atom in a real solid.[13] This approaches the fundamental limitation of the fixed node approximation, which introduces errors of this magnitude, as has been shown in atomic and molecular calculations. However, there are still developments required to have effective methods to carry out fixed node calculations with non-local potentials.[14, 15] The second conclusion is that promising results have been found using a new approach to "core/valence partitioning" that leads to "quasiparticle pseudopotentials." [18, 19] These have been derived for atoms using many-body Green's function methods to determine the self-energy of a valence electron due to exchange and correlation with the core. Tests have shown the new potentials are transferable and predict many different states of atoms and binding of the Na dimer more accurately than potentials derived from single body methods like Hartree-Fock or LDA. Together these results are promising for accurate *an initio* calculations on molecules and solids.

ACKNOWLEDGEMENTS

We acknowledge many useful discussions with J. Yu, G. Ortiz and many other colleagues. This project was supported by the NSF grants DMR-91 17822 and DMR- 89 20538 and the U. S. Department of Energy under grant DE FG02-91ER4539. Computations were done at the National Center for Supercomputer Applications at the University of Illinois.

References

[1] D. M. Ceperley and M. H. Kalos, in Monte Carlo Methods in Statistical Physics, edited by K. Binder (Springer, Berlin, 1979); K. E. Schmidt and M. H. Kalos, in Monte Carlo Methods in Statistical Physics II, edited by K. Binder (Springer, Berlin, 1984).

[2] D. M. Ceperley, and B. Bernu, J. Chem. Phys. **89**, 6318(1988).

[3] D. M. Ceperley, G. V. Chester, and M. H. Kalos, Phys. Rev. B **16**, 3081(1977).

[4] S. Fahy, X. W. Wang and S. G. Louie, Phys. Rev. Lett. **61**, 1631(1988); Phys. Rev. **B42**, 3503(1990).

[5] D. M. Ceperley, Phys. Rev. B **18**, 3126(1978); D. M. Ceperley and B. J. Alder, Phys. Rev. Lett. **45**, 566(1980).

[6] P. J. Reynolds, D. M. Ceperley, B. J. Alder and W. A. Lester, J. Chem. Phys. **77**, 5593(1982); D. M. Ceperley and B. J. Alder, ibid. **81**, 5833(1984).

[7] Some authors distinguish between GFMC and DMC. Our DMC calculations use a short time approximation to the Green's function (see Ref. [6]), and we have tested that the time step is small enough that this is not an essential approximation.

[8] M. H. Kalos, preprint.

[9] D. M. Ceperley and B. J. Alder, Phys. Rev. **B36**, 2092(1987).

[10] G. B. Bachelet, D. M. Ceperley and M. G. B. Chiochetti, Phys. Rev. Lett. **62**, 2088(1989).

[11] D. R. Hamann, M. Schlüter and C. Chiang, Phys. Rev. Lett. **43**, 1494(1979); G. B. Bachelet, D. R. Hamann and M. Schlüter, Phys. Rev. **B26**, 4199(1982).

[12] W. M. C. Foulkes and M Schlüter, Phys. Rev. **B42**, 11505(1990).

[13] X. P. Li, D. M. Ceperley, and R. M. Martin, Phys. Rev. **44**, 10929 (1991).

[14] L. Mitáš, E. L. Shirley and D. M. Ceperley, J. Chem. Phys. **95**, 3467 (1991).

[15] L. Mitáš, in Computer Simulations in Condensed Matter Physics IV, ed. D. P. Landau, Springer (1992); earlier references on approaches to deal with core electrons and non-local potentials in QMC are given in this paper.

[16] V. Natoli, R. M. Martin, and D. M. Ceperley, submitted for publication.

[17] X. W. Wang , S. Fahy, and S. G. Louie, Phys. Rev. Lett. **65**, 2414(1990).

[18] E. L. Shirley,thesis, University of Illinois, 1991; E. L. Shirley and R. M. Martin, to be published.

[19] E. L. Shirley, L. Mitáš, and R. M. Martin, Phys. Rev. **44**, 3395 (1991).

[20] Y. Kwon, D. M. Ceperley, and R. M. Martin, to be published.

[21] L. Hedin and S. Lundquist, Solid State Physics, vol. 23, p.1 (Academic Press, New York, 1969).

[22] W. Muller, J. Flesch, and W. Meyer, J. Chem. Phys. **80**, 3297 (1982); W. Muller, and W. Meyer, J. Chem. Phys. **80**, 3311 (1982).

[23] X. P. Li, R. J. Needs, R. M. Martin, and D. M. Ceperley, Phys. Rev. **45**, 6124 (1992).

[24] E. Krotscheck and W. Kohn, Phys. Rev. Lett. **57**, 862 (1986).

[25] Z. Y. Zhang, D. C. Langreth, and J. Perdew, Phys. Rev. **41**, 5674 (1990).

Application of GGA to a Study of Structural and Magnetic Properties of 3*d* Transition Metals

T. Asada

Department of General Education, College of Engineering, Shizuoka University, Hamamatsu 432, Japan

Abstract. Structural and magnetic properties of bcc, fcc, and hcp Fe and Mn are calculated by use of the generalized gradient approximation (GGA) proposed by Perdew and Wang. For both metals calculated results reproduce most of relevant cohesive properties found experimentally. For example, the pressure induced hcp-bcc phase transition of Fe can be successfully described and an antiferromagnetic ground state is predicted for the fcc structure of Mn with lattice constant and magnetic moment in good agreement with the experiment. Investigation of finite temperature behaviors of bcc, fcc, and hcp Mn by use of quasi-harmonic approximation removes apparent difficulties observed in the calculated results when compared with the experimental facts at high temperature.

1. Introduction

The generalized gradient approximation (GGA) proposed by Perdew and Wang [1] and Perdew [1] has been tested by several workers. It was shown that the new functional corrected most of the errors in the exchange and correlation energies by local density approximation (LDA) in the density functional theory [1], and that it could predict correctly the bcc-ferromagnetic (FM) phase for the ground state of Fe [2]. It was also shwon that for C, Al and Si GGA brought about sizable improvements in cohesive energies [3]. However it was also found that the results by GGA were sometimes inferior to those by LDA, especially for 4*d* and 5*d* metals [4]. Therefore the reputaion of GGA seems somehow controversial in the present stage. However, the improvements achieved for cohesive properties of 3*d* metals and lighter elements should be properly appreciated. The purpose of this paper is to clarify the extent to which GGA can claim its success for Fe by a thorough search on the fundamental structures of lattice (bcc, fcc and ideal hcp) and magnetic phases (nonmagnetic (NM), FM, and antiferromagnetic (AF)) for Fe and Mn and thus to contribute to an assessment of GGA and its refinement.

For the magnetic structure of AF phase we assume a spin ordering reported for fcc-AF Mn [5], where the magnetic moments in a (001) plane are parallel to each other but antiparallel to those in alternate sheets. No magnetically induced lattice distortions are taken into account. We will show below that the introduction of modulations in magnetic and/or lattice structure would not alter conclusions to be stated below.

2. Method of Calculation

The results presented here were calculated by the semi-relativistic linear muffin-tin orbital (LMTO) method [6,7] in the atomic sphere approximation (ASA) with the so-called com-

Springer Series in Solid-State Sciences, Volume 114
Interatomic Potential and Structural Stability Eds.: K. Terakura and H. Akai

bined correction. The basis consisted of *spdf*-LMTO's. We employed the linear tetrahedron method for sampling k-points and each eigenstate was weighted by Fermi distribution function with a broadening factor of 1 mRy. For the NM or FM phase of bcc (fcc) lattice, we employed a unit cell which contained two atoms, *i.e.*, simple cubic (sc) (body-centered tetragonal (bct)) unit cell, the one for the AF phase, in order to avoid possible errors of several mRy introduced in handling different unit cells for different magnetic phases. The numbers of irreducible k-points were 220 (165) for sc and 216 (140) for bct, and 180 (140) for hcp lattice of Fe (Mn). A common lower cut-off radii for the radial integrations were chosen for all calculations since the total energy by GGA does not converge with respect to the cut-off radius. The Ceperley-Alder functional was used for LDA. Since the value of the magnetic moment could foat in the present self-consistent field (SCF) scheme, the iterations, sometimes for the case of possible appearance of two spin states, became very intractable for either one or both of two spin states. Branches shown in this paper are those whose convergences were strictly achieved. The magnetic states were identified as NM if values of the magnetic moment, μ, were effectively zero (less than $\sim$0.05 μ_B). Small spin states often went finally to NM after careful and sufficient iterations.

Quantities such as equilibrium Wigner-Seitz radius, r_{WS}, cohesive energy, E_{coh}, bulk modulus, B_0, ans its pressure derivative, B_0', were obtained by least-mean-squares fitting of the calculated total energies to the Murnaghan equation of state [8],

$$E(V) = E(V_0) + \frac{B_0 V}{B_0'(B_0' - 1)}\left\{B_0'(1 - \frac{V_0}{V}) + (\frac{V_0}{V})^{B_0'} - 1\right\} , \tag{1}$$

where E, V, and V_0 are the total energy, an atomic volume and its equilibrium value, respectively.

3. Calculated Results and Discussions

A. Fe

Relevant experimental facts to be compared with the present calculation are summarized as follows:

(1) Under zero pressure the ground state of Fe is bcc-FM, with r_{WS} = 2.66 a.u., B = 1.68 Mbar, E_{coh} = 0.317Ry and μ = 2.12 μ_B [9]. The next lowest phase will be the hcp phase with the fcc (NM or AF) phase above it. [10]

(2) For the fcc phase, two experimental observations under bulk environments have been made on precipitates, *i.e.*, one in Cu is AF with r_{WS}=2.65 a.u. and μ $\sim$0.7 μ_B [11], and the other in CuAu alloys is FM with r_{WS} $\sim$2.78 a.u. [12].

(3) Under an increasing pressure a bcc-FM to hcp-NM phase transition occurs at a pressure of $\sim$140 kbar at low temperature [13]. At the transition point r_{WS} is $\sim$2.61 a.u. in the bcc-FM phase and $\sim$2.56 a.u. in the hcp-NM phase with c/a $\sim$1.60 [14].

First we show in Fig.1, the total energy, E_{tot}, and μ, as a function of r_{WS}, calculated by LDA. The overall features are essentially the same as those by earlier calculations [15,16]. The high-spin (HS) branch for the fcc-FM phase was not obtained in the region searched. Except for the bcc-FM, the stable HS branches obtained collapse at a certain critical point in decreasing r_{WS}. Only one low-spin (LS) solution for fcc-FM was obtained. The difficulties in obtaining both HS and LS solutions in the fcc-FM phase in the range searched, the absence of LS branch in the hcp-FM phase and the behavior of

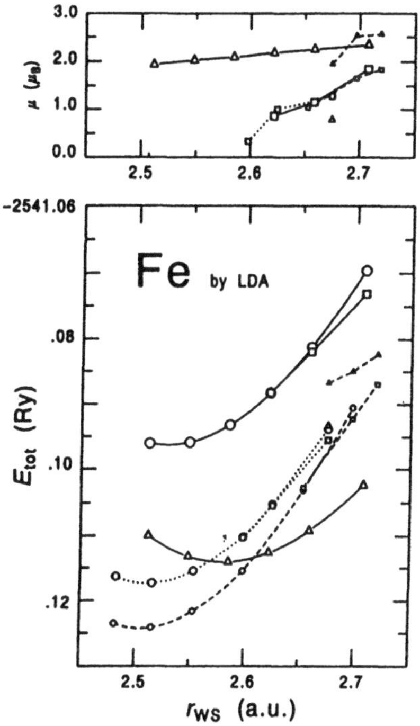

Fig.1 E_{tot}, and μ, of Fe as a function of r_{WS}, calculated with Ceperley-Alder LDA functional. The zero-point energy is not included. The solid curve corresponds to the bcc structure, the dotted to the fcc, and the dashed to the hcp. The circles indicate NM, the triangles FM, and the squares AF.

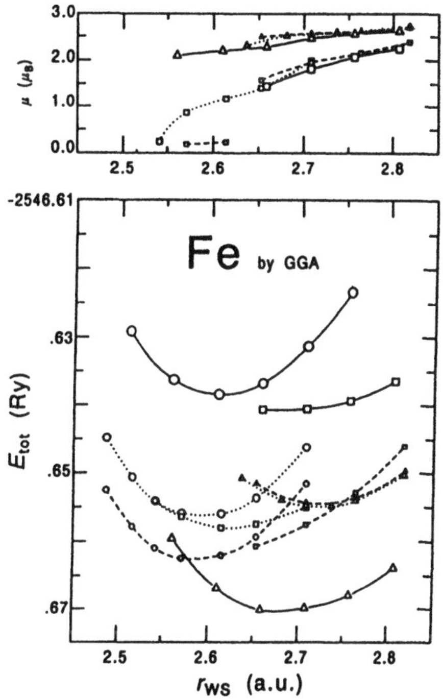

Fig.2 E_{tot}, and μ, of Fe as a function of r_{WS}, calculated by GGA. The zero-point energy is not included. The solid curve corresponds to the bcc structure, the dotted to the fcc, and the dashed to the hcp. The circles indicate NM, the triangles FM, and the squares AF.

μ change in the fcc-AF phase can be understood by the help of earlier calculations by the fixed-spin moment method [16]. Though LDA gives rather correctly mutual locations of the NM phases of three lattice structures in the $r_{WS} - E_{tot}$ diagram, it is difficult to seek agreements with the experimental facts stated above, because LDA, along with its disposition for predicting smaller lattice constants, produces magnetic energy gains insufficient or too small. From Fig.1 we observe that LDA predicts a false ground state of the hcp-NM with no possibility of the pressure-induced hcp-bcc phase transition nor of the existence of any stable AF or FM phases in the required r_{WS} range of the fcc lattice.

Fig.2 shows the GGA results of E_{tot} and μ as a function of r_{WS}. The HS branches were stably obtained for FM and AF phases of all three structures. The LS branch was obtained only for the hcp-AF phase. Comparing Fig.2 with Fig.1, we observe that GGA has an inclination to produce larger interatomic distances and that the magnetic energy gain by GGA is significantly larger than the one by LDA. Owing to these features, the total energy diagram shifts toward larger r_{WS} and the total energy of magnetic phases lowers a great deal by formation of magnetic moment. As seen in Fig.2, the ground state by GGA is correctly bcc-FM and corresponding equilibrium values are $r_{WS} = 2.683$ a.u.,

Table 1: Comparisons of the calculated results of Fe with corresponding experimental results. The Ceperlay-Alder functional is used for LDA. Wigner-Seitz radius, r_{WS}, is in a.u., bulk modulus, B, in Mbar, cohesive energy, E_{coh}, in Ry, magnetic moment, μ, in μ_B, and pressure, P, in kbar

		Theory		Experiment
		LDA	GGA	
ground state		hcp-NM	bcc-FM	bcc-FM
bcc-FM	r_{WS}	2.58	2.68	2.66[a]
	B	2.40	1.58	1.68[a]
	E_{coh}	0.491	0.359	0.317[a]
	μ	2.10	2.32	2.12[a]
fcc-AF	r_{WS}	—	2.62	2.65[b]
	μ	—	1.21	0.7[b]
-FM	r_{WS}	—	2.72	2.78[c]
	μ	—	2.56	—
bcc-hcp transition	P	—	149	~140[d]
	$r_{WS}{}^{hcp}$	—	2.54	2.56[e]
	$r_{WS}{}^{bcc}$	—	2.62	2.61[e]

[a]Ref.9, [b]Ref.11, [c]Ref.12
[d]Ref.13, [e]Ref.14

$B = 1.58$ Mbar, and $\mu = 2.32\ \mu_B$ and E_{coh} is 0.359Ry. These values are essentially the same as those by Bagno *et al.* [2] and show great improvements on r_{WS}, B, and E_{coh}. The discrepancy in the magnetic moment seems to come from the present treatment by ASA and the inclusion of the non-spherical potential will amend most of the errors (See, *e.g.*, Singh *etal.* of Ref.3). As for the fcc lattice, Fig.2 predicts one stable AF phase with $\mu = 1.21\ \mu_B$ at $r_{WS} = 2.62$ a.u. and one FM with $\mu = 2.56\ \mu_B$ at $r_{WS} = 2.72$ a.u.. The values of r_{WS}'s, 2.65 a.u. and 2.78 a.u., observed on micro-precipitates in Cu [11] and CuAu alloys [12] seem to correspond to these equilibrium values. Their agreements seems to be rather good, though the theoretical magnetic moment, 1.21 μ_B, is too large compared with the experimental value of ~0.7 μ_B [13]. We next discuss the relative location between the total energy curves of bcc-FM and hcp-NM phases, which concerns the pressure induced hcp-bcc transition at low temperature. From Fig.2 we can expect the transition in the direction observed experimentally. The transition pressure is deduced to be 149 kbar and the critical r_{WS} on the bcc-FM phase is 2.62 a.u. and 2.54 a.u. on the hcp phase. The agreements of the pressure and the r_{WS}'s with the corresponding experimental values of ~140 kbar, 2.61 a.u. and 2.56 a.u. [13,14], respectively, seem quite satisfactory. Finally we mention some possibility that the electronic structure by GGA might be favorable for describing finite temperature behaviors of Fe. The total energy difference between the equilibrium values of fcc-AF and hcp-AF phases is 4.6mRy, and the one between those of hcp-AF and bcc-FM phases is 7.6mRy by the present calculation. These values correspond to 1.0mRy and 7.3mRy, respectively, which were used by Hasegawa and Pettifor [17] as energy separations at absolute zero temperature in their semiquantitative treatment of phase transition of Fe at finite temperature based on the spin fluctuation mechanism. Taking their success into account, the quantitative agreement in these values convinces us that GGA might provide a sound starting point for treating the finite temperature behaviors of Fe. Comparisons of the calculated results with the experimental ones are compiled in Table 1.

Experimentally the fcc-AF lattice was found to contract slightly along c axis by 0.17 % and to expand by 0.19 % in the c plane with a sinusoidal modulation of the atomic position [18]. Inclusion of such small distortions or other possible modulations in lattice and/or magnetic structures would affect little the above conclusions, since associated total energy change is estimated to be sufficiently small [19].

B. Mn

Mn occurs in four allotropic forms, *i.e.*, α, β, γ, and δ phases, with increasing temperature and at the atmospheric pressure [20]. The first two phases, α and β, have complex structures with 29 and 20 atoms per unit cell, respectively. The α phase exists up to ~1073 K. The γ (δ) phase is simple fcc (bcc) and occurs between ~1373 (1407) and 1407 (~1518) K. Since the structures of the first two phases are too complicated, our concern here is restricted mainly on the other two high temperature simple phases, γ(fcc) and δ(bcc). For the fcc structure it is known that the magnetic structure is AF with $\mu = 2.30$ μ_B, $r_{WS} = 2.752$ a.u. (at room temperature (r.t)) and the Néel temperature $T_N = 540$ K [21] and $r_{WS} = 2.853$ a.u. at 1373 K [20]. For bcc structure r_{WS} is 2.867 a.u. at 1413 K [20] and its magnetic structure has not been manifested yet.

Fig.3 shows the LDA results of E_{tot} and μ as a function of r_{WS}. The result for each of the three lattice structures is essentially the same as the corresponding one reported earlier [22], though no comparison is possible with respect to their relative stability since no such calculations as Fig.3 are available so far [23]. At r_{WS} ~2.59 a.u. around which all phases have their minimum energy, we obtained only two magnetic branches, bcc-FM and fcc-AF. The lowest energies for these magnetic solutions are almost degenerate to the NM

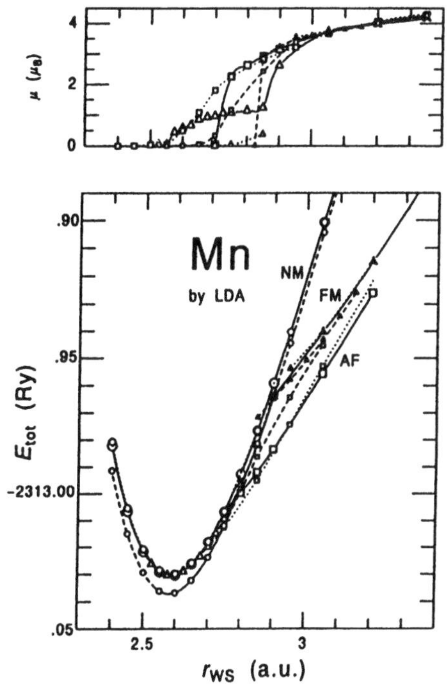

Fig.3 E_{tot}, and μ, of Mn as a function of r_{WS}, calculated with Ceperley-Alder LDA functional. The zero-point energy is not included. The solid curve corresponds to the bcc structure, the dotted to the fcc, and the dashed to the hcp. The circles indicate NM, the triangles FM, and the squares AF.

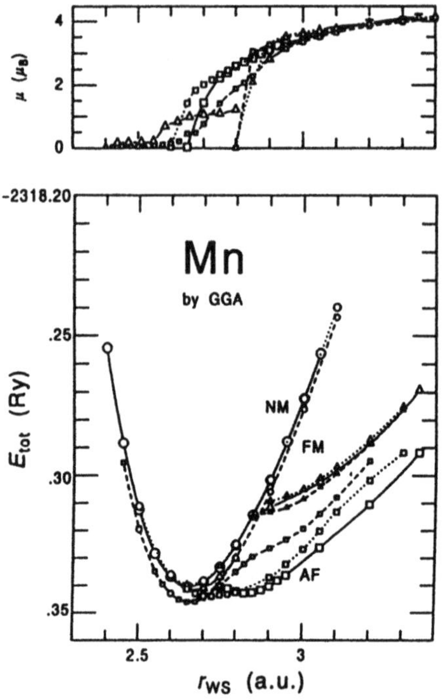

Fig.4 E_{tot}, and μ, of Mn as a function of r_{WS}, calculated by GGA. The zero-point energy is not included. The solid curve corresponds to the bcc structure, the dotted to the fcc, and the dashed to the hcp. The circles indicate NM, the triangles FM, and the squares AF.

ones within numerical accuracy of ~0.1 mRy. The lowest energy state among these three structures is hcp with 6.8 mRy difference to the next lowest state of bcc. The difference between bcc and fcc minima is 0.5 mRy. Compared with experimental results, LDA gives too small lattice constants and cannot predict unambiguously the magnetic structure of the ground state. LDA would predict the hcp structure, rather than fcc or bcc, for the high temperature phase even when we took the effect of vibrational entropy into account by the way we show below.

In Fig.4 we show the GGA counterpart of Fig.3. Compared with Fig.3, we perceive that the magnetic energy gain brings about sizable effects on the total energy behaviors of the magnetic branches. Findings on Fig.4 are as follows:

(1) For the hcp structure, the ground state is AF with r_{WS} of 2.663 a.u. and μ of 1.52 μ_B, which is lower than the NM phase marginally by 0.4 mRy. FM solutions are found only for r_{WS} greater than ~2.85 a.u..

(2) For the fcc structure, the ground state is AF with r_{WS} of 2.740 a.u. and μ of 2.30 μ_B, which are in good agreement with the experimental values given above, though the agreement in μ would be partly due to the treatment by ASA as was mentioned in the above subsection. The total energy difference with that of NM (or almost degenerate FM) is 5.4 mRy, which definitely supports the AF ground state for this structure.

(3) For the bcc structure, the ground state is AF with $r_{WS} = 2.827$ a.u. and $\mu = 2.75$ μ_B. The FM minimum is located 1.4 mRy above the AF with $r_{WS} = 2.68$ a.u. and $\mu =$ 0.97 μ_B. The equilibrium of the NM phase has the highest total energy, 0.9 mRy above that of FM.

(4) The most stable phase among those investigated is hcp-AF. The total energy difference between hcp-AF and the next lowest fcc-AF is 2.5 mRy, and 1.8 mRy between fcc-AF and the highest bcc-AF. We observe too large a difference between r_{WS}'s of fcc-

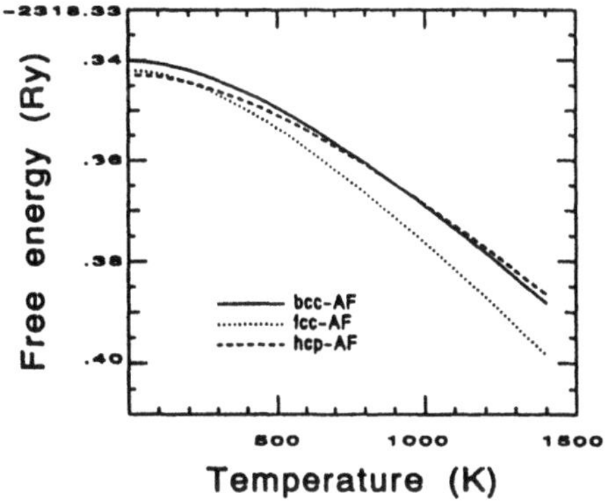

Fig.5 Free energy change of bcc-, fcc-, and hcp-AF states shown in Fig.4 as a function of temperature.

Table 2: Quantities for Mn deduced from Fig.4 and experimental values. The total energy minimum, $E_{\rm min}$, is in Ry, bulk modulus, B, in Mbar, Wigner-Seitz radius, $r_{\rm WS}$, in a.u. and magnetic moment, μ, in $\mu_{\rm B}$. γ means Gruneisen constant.

structure	ground state	$E_{\rm min}$	B	γ	$r_{\rm WS}$	μ
bcc	AF	-2318.340	1.50	0.44	2.832	2.76
fcc	AF	-2318.342	0.86	1.42	2.737	2.25
(300 K)					2.746	2.31
(exp. r.t.)	AF				2.752[a]	2.30[a]
hcp	AF	-2318.343	2.42	1.29	2.663	1.52

[a] Ref.21

and bcc-AF, when compared with their high temperature values which are almost identical with each other.

To know how these equilibrium states evolve with temperature, we investigate the consequence of the vibrational entropy by using the quasi-harmonic approximation [24]. By the approximation the free energy, F, as a function of $r_{\rm WS}$ and temperature T can be expressed as follows:

$$F(r_{\rm WS},T) = E_{\rm tot}(r_{\rm WS}) - k_{\rm B}T\Big\{D(\Theta_{\rm D}) - 3\ln(1-e^{-\Theta_{\rm D}/T})\Big\} + \frac{9}{8}k_{\rm B}\Theta_{\rm D} \ , \qquad (2)$$

where $\Theta_{\rm D}$ is the Debye temperature and $D(x)$ is the Debye function. The last term of Eq.(2) represents the zero point energy. In Eq.(2), the $r_{\rm WS}$ dependence of the phonon part enters by way of $\Theta_{\rm D}$ through the relation

$$\Theta_{\rm D}/(\Theta_{\rm D})_0 = ((r_{\rm WS})_0/r_{\rm WS})^{3\gamma} \ , \qquad (3)$$

where the suffix 0 stresses that the quantities are for the equilibrium. The Gruneisen constant, γ, is related to B_0' used in the Murnaghan equation of state, Eq.(1), through the relation

$$\gamma_0 = \frac{B_0' - 1}{2} \ , \qquad (4)$$

which should hold at rather low temperature [24].

Fig.5 show the change of the free energy, F, evaluated by use of Eq.(2) but with quantities listed in Table 2, which are deduced from the rigid lattice total energy, $E_{\rm tot}$, of

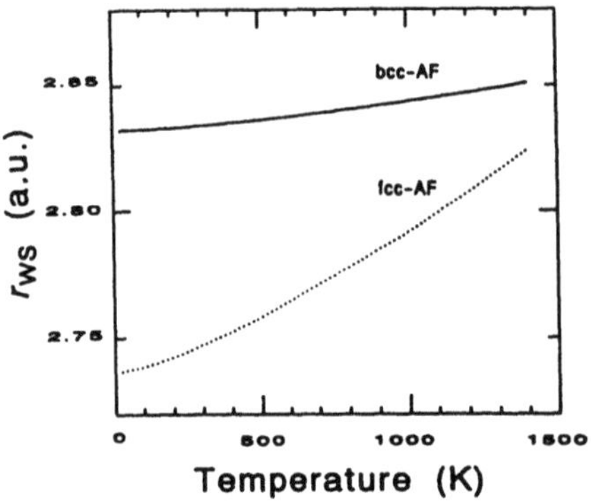

Fig.6 r_{WS} change as a function of temperature.

Fig.4 [25]. We observe in Fig.5 that the hcp-AF phase is replaced by the fcc-AF phase at ~200 K, which comes from the much larger value of B_0 of the hcp-AF phase compared with that of the fcc-AF phase. Therefore, since the α phase free energy must surely be lower than this hcp value at low temperature, this hcp phase, if not stabilized by alloying, would not be realized experimentally without some structure-forcing technique such as molecular-beam-epitaxy. This situation meets the experimental facts stated above. Experimentally the fcc phase is replaced at 1407 K. The difference in the free energies of fcc and bcc at around 1400 K is almost 10 mRy in Fig.5, which is too large to fill up by the difference in the magnetic entropies of two phases. It seems to require more careful treatment of the vibrational effect to discuss the behaviour at such high temperature. We show in Fig.6 the behavior of r_{WS} at finite temperature, which should be taken only qualitatively at rather high temperature as for Fig.5. Since the Gruneisen constant for the bcc-AF phase is very small as shown in Table 2, r_{WS} for this phase increases only slightly with temperature. This behavior makes a clear contrast with the one of the fcc-AF phase. Fig.6 would remove an apparent difficulty seen in the difference between calculated values of r_{WS}'s for fcc- and bcc-AF phases, which is too large in comparison with their almost identical experimental values at high temperature.

Here we briefly mention the consequence of magnetically induced distortion observed in the fcc-AF phase [21]. Below T_N the fcc lattice shows a tetragonal distortion of about a 6 % contraction along [001] axis, which is considerably larger than the case of fcc Fe [18]. The energy lowering by this distortion was calculated to be ~2 mRy [19], which is of the same order of total energy difference between the three participating AF phases in Fig.4. The energy lowering thus might introduce a delicate situation to the relative location of these three ground state energies. However it would not cause any difficulty in the above discussion of the high temperature behavior, since for the present case quantities such as B_0 and B_0', of each ground state determine decisively the behaviors at high temperature as we have just demonstrated.

C. Some characteristic features of GGA

So far we have shown that GGA can reproduce satisfactorily most of the relevant facts related to the cohesive properties of Fe and Mn. Therefore we believe the present calculation firmly substantiates the success of GGA shown for Fe by earlier workers.

Comparing GGA results with those of LDA, we can list characteristic features of GGA as follows:

(1) It has an inclination to produce larger interatomic distances.
(2) Its magnetic energy gain is significantly larger than the one by LDA.

These two features come mostly from the functional form of the GGA functional. It can be clearly shown by a perturbative evaluation of the GGA exchange-correlation energy (say, of Fe), in which the use of an LDA SCF electron density in GGA recovers most of the total energy difference of ~6 Ry between those of LDA and GGA with remaining small difference (~7 mRy) to be amended by redistribution of electron density. Then the SCF electronic structure and dispersion of GGA could remain almost unchanged from those by LDA, which is the fact confirmed so far. The origin of the first feature can easily be traced to the functional form of the GGA exchange-correlation energy density: The GGA correction to the exchange (correlation) energy density is a negative (positive) and ever decreasing (increasing) function of the inhomogeneity, s, with $s \propto |\nabla\rho|/\rho^{4/3}$, where ρ is the electron density. Since the magnitude of the correction to the exchange part is several times larger than that to the correlation part, the total correction to the exchange-correlation energy density is a negative ever decreasing function of s. Since ρ decreases and $|\nabla\rho|$ increases for lattice expansion in general, the quantity s increases with the expansion, which leads to the first feature. The second feature, however, does not come straightforwardly from the functional form. We hope its generality will be confirmed numerically by further examples. It might be conceived that the feature might come from the enhancement of the magnetic moment. But it should be generally denied since the feature stands also for the GGA full-potential calculation, where the magnetic moment is reduced from the one by LDA; [See Singh *et al.* of Ref.2].

It should be mentioned that the Kohn-Sham exchange-correlation potential of GGA, or generally of gradient correction method, has $-1/r$ divergence with $r \to 0$. Though this divergence little affects integrated values such as the total energy, it seems to be fatal for treating the hyperfine field.

4. ConcludingRemarks

We have performed the total energy calculation as a function of lattice spacing for Fe and Mn by use of both LDA and GGA for three fundamental crystal structures, *i.e.*, bcc, fcc, and ideal hcp, and for magnetic structures of FM, AF, and NM. We have found that GGA works quite well for these magnetic substances: (1) For Fe, the ground state, the existence of stable AF and FM phases of fcc structure and the hcp-bcc pressure induced phase transition have been successfully reproduced with associated quantities in good agreemnt with the experimental results. (2) For Mn, the AF phase has been found to be the ground state of the fcc structure with the lattice spacing and magnetic moment in good agreement with the experiment. The investigation of finite temperature evolution of the ground states by the quasi-harmonic approximation has found that the difficulties observed in the calculated results of Mn would cause no troubles in the high temperature behavior.

References

1 J.P.Perdew and Y.Wang, Phys. Rev. B **33**, 8800 (1986); J.P.Perdew, *ibid.* **33**, 8822 (1986).

2 P.Bagno, O.Jepsen, and O.Gunnarsson, Phys. Rev. B **40**, 1997 (1989); D.J.Singh, W.E.Pickett, and H.Krakauer, *ibid.* **43**, 11628 (1991); T.C.Leung, C.T.Chan, and B.N.Harmon, *ibid.* **44**, 2923 (1991).

3 X.J.Kong, C.T.Chan, K.M.Ho, and Y.Y.Ye, Phys. Rev. B **44**, 2923 (1991).

4 B.Barbiellini, E.G.Moroni, and T.Jarlborg, J.Phys.: Condens.Matter **2**, 7597 (1990).

5 D.Meneghetti and S.S.Sidhu, Phys.Rev. **105**, 130 (1957); S.C.Abrahams, L.Guttman, and J.S.Kasper, *ibid.* **127**, 2052 (1962).

6 O.K.Andersen, Phys. Rev. B **12**, 3060 (1975); H.L.Skriver, *TheLMTOMethod* (Springer, Berlin, 1984).

7 D.D.Koelling and B.N.Harmon, J.Phys.C: Solid State Phys. **10** 3107 (1977).

8 F.D.Murnaghan, Proc.Nat.Acad.Sci. U.S.A. **3**, 244 (1944).

9 H.Danan, A.Harr, and A.J.P.Meyer, J. Appl. Phys. **39**, 669 (1968); W.B.Pearson, *Handbook of Lattice Spacings and Structures of Metals and Alloys* (Pergamon Press, Oxford, 1964).

10 This sequence is concluded from the pressure–volume phase diagram of Fe by recognizing that the appearance of the bcc-FM phase is due to its large magnetic energy gain.

11 L.Kaufman, E.V.Clougherty and R.J.Weiss, Acta Met.**11**, 323 (1961).

12 D.L.Williamson, S.Bukshpan, and R.Ingalls, Phys.Rev. B **6**, 4194 (1972).

13 C.Abrahams, L.Guttman, and J.S.Kasper, Phys.Rev.**127**, 2052 (1962); G.Johanson, M.B.McGirr, and D.A.Wheeler, Phys.Rev. B **1** 3208 (1970).

14 U.Gonser, K.Krishel, and S.Nasu, J.Magn.Magn.Mater.**15-18**, 1145 (1980).

15 C.S.Wang, B.M.Klein, and H.Krakauer, Phys.Rev.Lett. **54**, 1852 (1985); J.Kübler, Phys.Lett. **81A**, 81 (1981); H.J.F.Jansen, K.B.Hathaway, and A.J.Freeman, Phys.Rev. B **30** 6177 (1984).

16 V.L.Moruzzi, P.M.Marcus, K.Schwarz, and P.Mohn, Phys.Rev. B **34** 1784 (1986); V.L.Moruzzi, P.M.Marcus, and J.Kübler, *ibid.* **39** 6957 (1989); M.Podgórny and J.Goniakowski, *ibid.* **42**, 6683 (1990).

17 H.Hasegawa and D.G.Pettifor, Phys.Rev.Lett. **50**, 130 (1983).

18 Y.Tsunoda and N.Kunitomi, J. Phys. F: Met. Phys. **18**, 1405 (1988).

19 T.Oguchi and A.J.Freeman, J.Magn.Magn.Mater. **46** L1 (1984).

20 R.W.G.Wyckoff, *CrystalStructure*, vol.1 (Wiley, New York, 1963) p.50.

21 Y.Endoh and Y.Ishikawa, J.Phys.Soc.Japan **30**, 1614 (1971).

22 J.L.Fry, Y.Z.Zhao, N.E.Brener, G.Fuster, and J.Callaway, Phys.Rev.B **36**, 868 (1987); V.L.Moruzzi, P.M.Marcus, and P.C.Pattnaik, *ibid.* **37**, 8003 (1988); S.Fujii, S. Ishida, and S.Asano, J.Phys.Soc.Japan **60**, 1193 (1991); G.Fuster, N.E.Brener, J.Callaway, J.L.Fry, Y.Z.Zhao, and D.A.Papaconstantopoulos, Phys.Rev.B **38**, 423 (1988); V.L.Moruzzi, P.M.Marcus, and J.Kübler, *ibid.* **39**, 6957 (1989); M.Podgórny and J.Goniakowski, *ibid.* **42**, 6683 (1990).

23 The stability of Mn was dealt with in trend discussions of relative stability based on force theorem calculations. See, *e.g.*, H.Skriver, Phys.Rev.B **31**, 1909 (1985).

24 V.L.Morruzi, J,F,Janak, and K.Schwarz, Phys.Rev.B **37**, 790 (1988).

25 The temperature dependence of F and r_{WS} to be shown in Figs.5 and 6, respectively, will be enhanced if we use B_0 and B_0' to be calculated by Eq.(2).

Metal-Insulator Transition in Ni Compounds

M. Takahashi[1] *and J. Kanamori*[2]

[1]Department of Physics, Osaka University, Toyonaka, Osaka 560, Japan
[2]Osaka University, Suita, Osaka 565, Japan

The metal-insulator transition in Ni compounds is discussed on the basis of two calculations which take account of electron correlation. The first calculation which assumes a linear model of a finite size simulating a compound NiX with X representing *p*-valence atom such as O and S determines rigorously the wave function of the ground state by use of the Lanczos method. The second one which adopts a realistic three dimensional model uses the Local Ansatz approach to take into account the correlation effect. It is concluded that a Ni atom in the metallic state near the transition keeps very likely a well developed magnetic moment. Correlation between neighboring Ni magnetic moments, however, will become much smaller than that in the insulating phase. The results are consistent with the photoemission experiment which observes quite a small change in the spectra across the transition.

1. Introduction

NiS is an antiferromagnetic insulator with the atomic moment of about $1.7\mu_B$ at low temperatures. When temperature or pressure is increased, it undergoes a phase transition into a metallic phase which is not antiferromagnetic[1]. The photoemission experiment on NiS made by Fujimori et al[2], on the other hand, yields the result that overall spectra do not change appreciably at the metal-insulator transition, suggesting that the local electronic structure does not change drastically contrary to the expectation based on the band structure calculation[3].

The purpose of the present study is to investigate the role of the electron correlation in the metal-insulator transition in Ni compounds where two d holes will occupy $e_g d$ orbitals of Ni atoms in the ionic limit. NiO was regarded to be a classical example of the Mott-Hubbard insulator; the discussion which focused attention on Ni ions supposed that the energy cost U_d required to produce Ni(d^7) and Ni(d^9) from two Ni(d^8) was big enough to suppress the itineracy of d electrons. However, Fujimori[4] and Sawatzky and Allen[5] have pointed out that the charge transfer en-

Springer Series in Solid-State Sciences, Volume 114
Interatomic Potential and Structural Stability Eds.: K. Terakura and H. Akai

ergy Δ involved in the process $\text{Ni}(d^8) \rightarrow \text{Ni}(d^9)\underline{\text{L}}$ with $\underline{\text{L}}$ representing a hole in the ligand is a key factor in the metal-insulator transition of Ni compounds, being smaller than U_d. Thus we need to adopt a model which takes into account not only Ni d states but also valence orbitals of ligand atoms in the discussion.

We present first the results of a rigorous determination of the ground state of a model Hamiltonian including the electron-electron interaction by use of the Lanczos method; we look for the change of the ground state wave function corresponding to the metal-insulator transition by varying a parameter representing the charge transfer energy. We conclude that Ni atoms keep a fairly large magnetic moment in a metallic phase neighboring the insulating antiferromagnetic phase; the interatomic correlation between magnetic moments of adjacent Ni atoms, however, is drastically reduced in this new phase. We calculate also the photoemission and BIS spectra. The calculated photoemission spectra do not change much at the transition which is consistent with experiment. The calculated BIS spectra of the new phase show that the phase has no gap at the Fermi energy, indicating a metallic nature of the ground state. We reported in a previous publication[6] the calculations except for the BIS spectra.

Since the above calculation assumes a simplified linear chain model, we have carried out also a calculation on a more realistic model through the Local Ansatz approach originally proposed by Fulde[7]. Although the Local Ansatz calculation has not been completed yet, we give here a preliminary report on the results which nevertheless support the previous conclusions.

In section 2 we describe the models and the methods of calculation. Results of the calculation for the linear chain model and those based on the Local Ansatz(LA) approach to a three dimensional model are presented in section 3 and 4 respectively. Section 5 is devoted to concluding discussions.

2. Model and calculations

We adopt the hole picture hereafter. A linear chain of Ni-X-Ni-X with X representing a ligand atom is assumed in the first calculation. Furthermore, we confine ourselves to the e_g orbitals of Ni and intervening $p\sigma$ ones of X in the tight binding scheme. For simulating the cubic symmetry of the real three dimensional crystal, the electron transfer is assumed to take place between $p\sigma$ and either of the $d\gamma$ orbitals with the same magnitude of the matrix element denoted by t. The intra-atomic coulomb interaction among the two d orbitals is taken into account fully under the assumption of the e_g symmetry. The intra-atomic coulomb interaction in $p\sigma$ of the ligand atom is also taken into account. Thus the model is described by the Hamiltonian

$$
\begin{aligned}
H = & \sum_{(\mathrm{NiL})n.n.,\sigma} (t\, C^{\dagger}_{d\sigma} C_{p\sigma} + h.c.) + \sum_{\mathrm{L}} \epsilon_p (n_{p\uparrow} + n_{p\downarrow}) \\
& + \sum_{\mathrm{L}} U_p n_{p\uparrow} n_{p\downarrow} + \sum_{\mathrm{Ni}} \sum_{\alpha\beta\gamma\delta} (\alpha\beta\|\gamma\delta) C^{\dagger}_{\alpha} C^{\dagger}_{\beta} C_{\delta} C_{\gamma}
\end{aligned} , \qquad (1)
$$

where $\alpha, \beta, \gamma, \delta$ denote a hole state in the $d\gamma$ orbitals including spin, ϵ_p denotes the hole energy of a p level measured from the d levels and $(\alpha\beta\|\gamma\delta)$ denotes the matrix elements of the intra-atomic coulomb interaction whose nonvanishing components will be expressed in terms of the coulomb integrals $J_{uu} = J_{vv}$,J_{uv} and exchange integral K_{uv} with u and v representing the $d\gamma$ orbitals. In our calculation the parameters representing the coulomb interaction are fixed as J_{uu}:J_{uv}:K_{uv}:U_p = 5:3:1:3. The parameter U_d characterizing the magnitude of the intra-atomic coulomb interaction of Ni is defined by $U_d = J_{uu} + K_{uv}$ which is equal to the ground state energy difference $E(d^9) + E(d^7) - 2E(d^8)$. The charge transfer energy Δ is defined by $\Delta = \epsilon_p - J_{uv} + K_{uv}$ which is equal to the ground state energy difference $E(d^9\underline{\mathrm{L}}) - E(d^8)$. For comparison we have carried out a calculation simulating a Cu compound adopting the model where only a $d\gamma$ orbital is assigned to each cation.

We calculate the ground state for the system of two holes per Ni atom. with the periodic boundary condition by use of the Lanczos method[8]. The calculation reported here is for the model containing 4 Ni sites, which is of the maximum size among tractable ones. We diagonalize the hamiltonian in the subspace, where the z-component of the total spin of the system is zero with four up spin holes and four down spin ones. In this case, where 8 $d\gamma$ orbitals and 4 $p\sigma$ ones are considered, the size of the Hilbert space is of 245025 states. Varying the charge transfer energy Δ, we study the change of ground state quantities such as the spin correlations. We expect that for large positive Δ each Ni atom accommodates two holes and has a large magnetic moment and for negative Δ with large absolute magnitude each ligand atom accommodate two holes in the present model. Since we assume for each ligand atom only one orbital in the present model, the latter case is not very meaningful. Thus we are mainly interested in the range of Δ which allows a small number of holes in the ligand atoms. In addition to the ground state properties we study the change of the photoelectron spectroscopy spectra using the same model with varying Δ.

In the Local Ansatz approach, we use a realistic three dimensional model where the Ni atoms and the ligand atoms constitute a rock salt structure and two $d\gamma$ orbitals are assigned to Ni atom and three $p(p_x,p_y,p_z)$

orbitals are assigned to ligand atom. The model is described by the following Hamiltonian.

$$\begin{aligned} H = &\sum(t_{pd}C^{\dagger}_{pi}C_{dj} + h.c.) + \sum(t_{pp}C^{\dagger}_{pi}C_{dj} + h.c.) \\ &+ \sum(t_{dd}C^{\dagger}_{di}C_{dj} + h.c.) + \sum \epsilon_p(n_{px} + n_{py} + n_{pz}) \\ &+ \sum(\alpha\beta\|\gamma\delta)C^{\dagger}_{\alpha}C^{\dagger}_{\beta}C_{\delta}C_{\gamma} \end{aligned} \tag{2}$$

In this calculation we neglect the coulomb interaction in the ligand atom for convenience, since we want to concentrate on the intra atomic correlation effect at Ni atom.

In the local Ansatz approach one first expresses the Hartree-Fock ground state $|\psi_{\mathrm{HF}}>$ of the system into a linear combination of configurations described in terms of the atomic orbitals. The trial function for the ground state $|\psi>$ is constructed by modulating the linear combination as

$$|\psi> = \exp(\sum \eta_m O^{(m)}_{ij})|\psi_{\mathrm{HF}}> \tag{3}$$

where η_m is a set of variational parameters, i, j run over all sites. $O^{(m)}_{ij}$ is a set of local operators, the general form of $O^{(m)}_{ij}$ can be written as

$$O^{(m)}_{ij} = \begin{cases} O^{(1)} = n_{i\uparrow} \cdot n_{i\downarrow} & , \\ O^{(2)} = n_i \cdot n_j & \text{for } i \neq j, \\ O^{(3)} = \vec{s}_i \cdot \vec{s}_j & \text{for } i \neq j. \end{cases} \tag{4}$$

For given local operators, the parameters (η_m) are determined by minimizing the ground state energy per site

$$E_G = \frac{1}{N}\frac{<\psi|H|\psi>}{<\psi|\psi>} \tag{5}$$

For calculating physical quantities, we use the Monte Carlo technique[9,10].

In the calculation we use the following approximation[11]. To calculate the average of a local operator such as $<\psi|h_i|\psi>$, where h_i is an onsite coulomb interaction operator or nearest neighbor spin correlation operator and so on, we take account of the projection operator related to the atoms located within a radius R from the atom to which the local operator h_i refers. We use $R = 1$ and $R = 2$ approximation where the unit is the distance to a neighboring ligand atom from a Ni atom.

3. Results of the linear chain model calculation

In the following we concentrate mostly on the calculation for the effective coulomb interaction $U_d = 6$; results for smaller U_d are similar. Figures 1a to 1c show the average of the total spin of each Ni atom, the spin correlation between neighboring Ni atoms and the hole density at the Ni atom and the ligand one for various Δ values, respectively; the energy is measured in the unit of the transfer integral t.

Figure 2 shows the phase diagram in Δ-U_d plane. In the region denoted by (I) each Ni atom has a large magnitude of spin larger than $1.6\mu_B$ with a strong antiferromagnetic correlation with neighboring Ni spins. We can assign this region to the insulating phase. Decreasing Δ, we go into the region denoted by (M). The boundary line between (M) and (I) is quite similar to that given by Zaanen et al.[12] on the basis of the impurity Anderson model. Following them we divide the region (I) into (I1) and (I2) which correspond to the Mott-Hubbard type insulator phase and the charge transfer one, respectively, though our calculation indicates no sharp distinction between them.

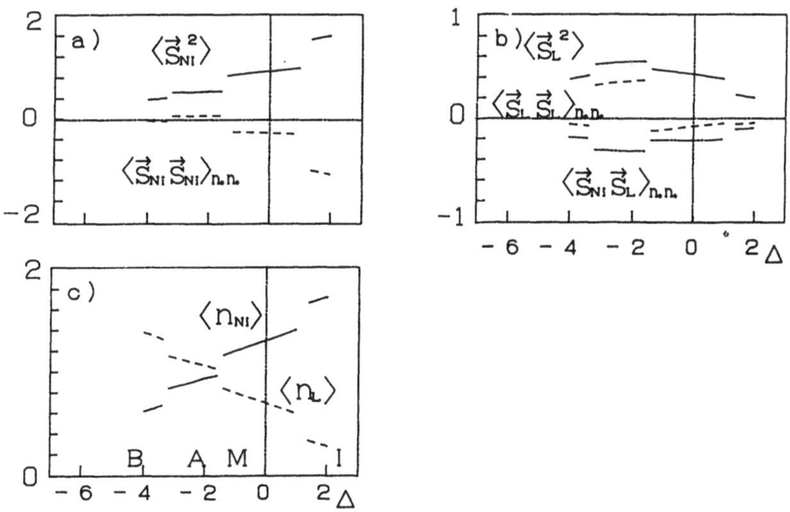

Fig. 1. Spin correlations and hole density at $U_d = 6$. (a) The solid line represents $< \vec{S}^2_{\mathrm{Ni}} >$ and the dotted line represents $< \vec{S}_{\mathrm{Ni}} \vec{S}_{\mathrm{Ni}} >_{n.n.}$ (b) The solid line represents $< \vec{S}^2_{\mathrm{L}} >$, the dotted line represents $< \vec{S}_{\mathrm{L}} \vec{S}_{\mathrm{L}} >_{n.n.}$ and two dotted broken line represents $< \vec{S}_{\mathrm{L}} \vec{S}_{\mathrm{Ni}} >_{n.n.}$ (c) The solid line represents $< n_{\mathrm{Ni}} >$ and the dotted line represents $< n_{\mathrm{L}} >$.

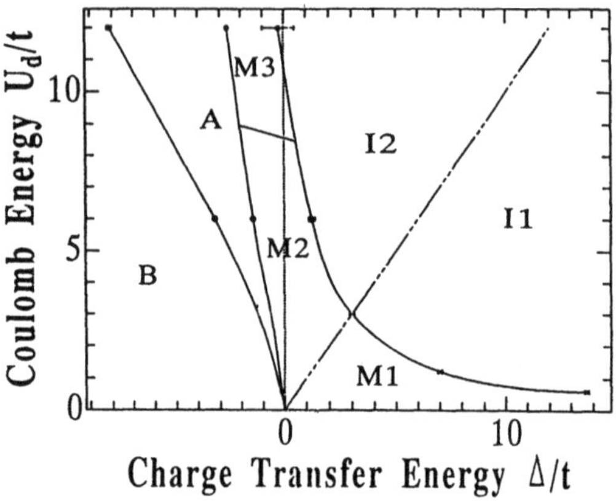

Fig. 2. A classification scheme based on the calculation described in the text. The region (I1) and (I2) correspond to the Mott-Hubbard type insulating phase and the charge transfer type insulating phase, respectively. The boundary between (M2) and (M3) is not known precisely. The details are explained in the text.

Across the (M)-(I)boundary line the ground state changes more than one times in a narrow range of Δ and the total spin takes a nonvanishing value in the transient range which, we believe, arises from the finite size effect. However, once inside the region (M) we reach the ground state with zero total spin which changes gradually with varying Δ. The subdivisions of the region (M) indicated in Fig.2 will be discussed below. The ground state is subject to more transitions by decreasing Δ further; the transitions are indicated by the boundary lines separating the regions (A) and (B) in Fig.2. The change of the ground state at these transitions is rather abrupt without a transient range.

In the following discussion we confine ourselves to the phase (M) and (I). In the region (M) each Ni atom accommodates more-than one hole and the total spin of a Ni atom is still high, corresponding to the magnetic moment larger than one e.g. $1.2\mu_B$ for U_d=6 near the transition between (I) and (M). The spin correlation between neighboring Ni atoms is, however, small compared to the region (I), whereas the antiferromagnetic spin correlation between Ni and the ligand increases. The Hund rule correlation between intra atomic orbitals in Ni atom decrease from about 0.3 to about 0.04 at the transition. The spin correlation between next nearest neighboring Ni atoms is even smaller in the subregions (M1) and (M2). In the region (M3) where a very large $U_d(\geq 8)$ is assumed, we find a strong antiferromagnetic correlation between the next nearest neighboring Ni spin. This is suggestive of a helical ordering with the rotating angle $\frac{\pi}{2}$ or a floating uudd structure, though it is difficult to pin-

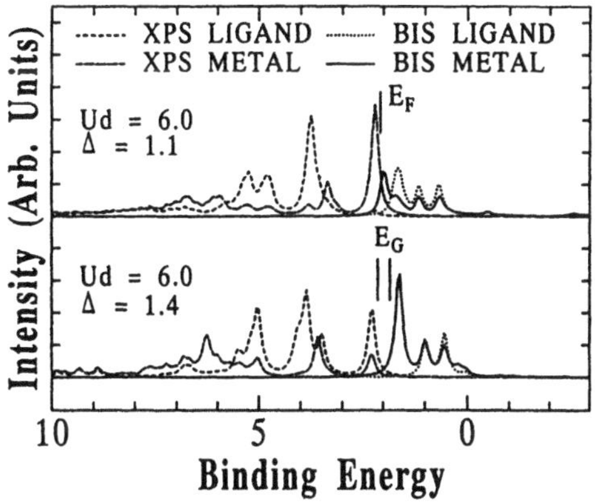

Fig. 3. The spectral densities of the Ni atom and of the ligand atom at $U_d = 6$. At $\Delta = 1.1$ the ground state belongs to (M2) and $\Delta = 1.4$ it belongs to (I2).

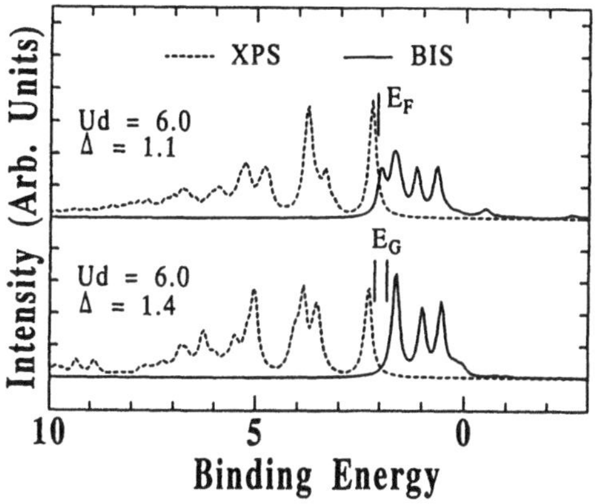

Fig. 4. The total spectra corresponding to photo electron spectroscopy spectra at $U_d = 6$

point the structure within the present calculation. The 4 spin interaction discussed by Yosida and Inagaki[13], for example, might be enhanced in the 4 Ni-atom cluster of the present calculation. We leave the possibility to future investigation.

Figure 3 shows the spectral density near the transition between (M) phase and (I) phase at $U_d = 6$. The solid line denotes the spectral density of Ni atom and the dotted line denotes that of ligand atom. We find that the location of the peaks, especially main peak, scarcely change across the transition between (I) and (M) which occurs between $\Delta = 1.4$ and 1.1 for $U_d = 6$. Figure 4 shows the total spectra including the

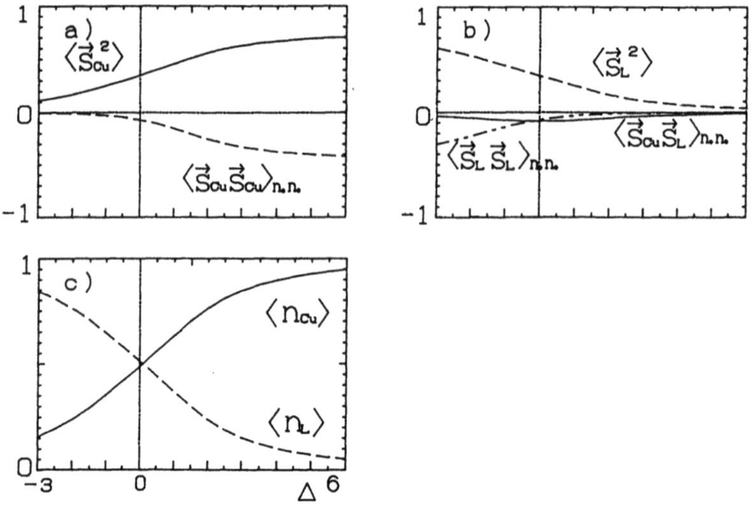

Fig. 5. Spin correlations and hole density at $U_d = 6$ for CuX case. (a) The solid line represents $< \vec{S}^2_{\rm Cu} >$ and the dotted line represents $< \vec{S}_{\rm Cu}\vec{S}_{\rm Cu} >_{n.n.}$. (b) The solid line represents $< \vec{S}^2_{\rm L} >$, the dotted line represents $< \vec{S}_{\rm L}\vec{S}_{\rm L} >_{n.n.}$ and two dotted broken line represents $< \vec{S}_{\rm L}\vec{S}_{\rm Cu} >_{n.n.}$. (c) The solid line represents $< n_{\rm Cu} >$ and the dotted line represents $< n_{\rm L} >$.

contribution from both Ni and ligand which should be compared with the photoemission spectra. The calculated BIS spectra are shown also to indicate the metallic nature of the ground state in the region (M).

In figure 5 we compare the present calculation on the Ni-X system with a similar calculation on the Cu-X system. The calculation for the latter case is carried out for the 6 Cu atom model. We can see that the system change gradually and the antiferromagnetic coupling between nearest neighbor Cu atoms decreases gradually with Δ. The abrupt decrease in the Ni case seems to show an interesting manifestation of the effect of the intra-atomic coulomb interaction.

4. Results for Local Approach

The table 1 shows the used parameters[14,15].

We use the paramagnetic (nonmagnetic) Hartree-Fock(PHF) solution as a starting wave function with the following projection operators.

$$O_i^{(m)} = \begin{cases} O^{(1)} = n_{i\uparrow} \cdot n_{i\downarrow} & \text{intra orbital } (i\text{: Ni atom}), \\ O^{(2)} = \vec{s}_{iu} \cdot \vec{s}_{iv} & \text{intra atom } (i\text{: Ni atom}). \end{cases} \qquad (7)$$

Table I. The used parameters. We renormalize them as $t_{pd\sigma} = 1$. The parameter ϵ_p represents the p level one body energy measured from the Ni d level.

ϵ_p valuable	$J_{uu} = J_{vv} = 5$	$J_{uv} = 3.744$	$K_{uv} = 0.628$
$t_{pp\sigma} = -0.4231$	$t_{pp\pi} = 0.1154$	$t_{pd\sigma} = 1$	$t_{pd\pi} = -0.4615$
$t_{dd\sigma} = 0.2271$	$t_{dd\pi} = -0.1034$	$t_{dd\delta} = 0.0109$	

When we use the antiferromagnetic Hartree-Fock(AHF) solution as the starting wavefunction, the projections do not change the state much. The AHF solution is obviously a good approximation to the ground state in the ionic state(the region (I)). On the other hand, we may suppose that the state obtained from the PHF may correspond to the ground state in the region (M).

In the calculation we check the convergency for the cluster size R mentioned in section 2. Because the variational principle breaks down due to the approximation, we might overestimate the energy gain from the paramagnetic Hartree-Fock energy. We first search the minimum energy point in the variational parameters space assuming $R = 1$ and recalculate the energy for $R = 2$ with the obtained parameter to check the convergency. We find that the energy difference between the case for $R = 1$ and $R = 2$ is smaller than about 20% of the energy gain in the range that the charge transfer energy is less than 4, however it is larger than about 25% at the large value of the charge transfer energy ($\Delta = 13$). Thus we suppose that in the small Δ range the energy converges roughly for R.

Figure 6 shows the total energy per unit cell where the solid line, the dotted line and dashed line represent the energy of paramagnetic Hartree-Fock wave function, the energy of the improved PHF wave function(LA-PHF) and the energy of the antiferromagnetic Hartree-Fock wave function(AHF). In the range where the charge transfer energy is less than about 4, the energy difference between the LA-PHF and AHF becomes very small. We can lower the energy of the LA-PHF further by using the PHF wave function corresponding to a larger charge transfer energy as a starting wave function, as is shown by the mark ($\times$) in Fig. 6. The projections increases the number of holes at Ni-sites and produce a magnetic moment there. The idea of adopting the PHF wave function corresponding to a larger Δ is to prepare a larger number of holes at Ni-sites at the start. Figure 7 shows the square of spin at Ni atom and the spin correlation between the next nearest neighbor Ni atoms. The square of spin at Ni atom for the LA-PHF wave function is larger than for the PHF wave

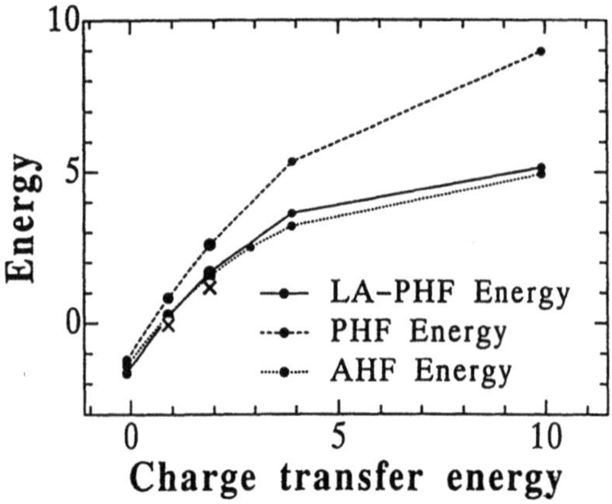

Fig. 6. Total energy per unit cell. The dashed line, the solid line and dotted line correspond to the paramagnetic Hartree Fock wave function, correlated wave function obtained from PHF wave function and the antiferro magnetic Hartree Fock wave function respectively.

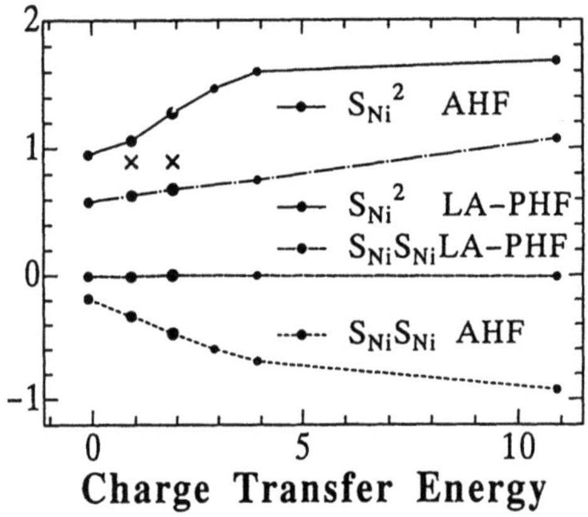

Fig. 7. Spin correlations. The points marked (×) represent $< \vec{S}^2_{\rm Ni} >$ obtained by using the PHF wave function at $\Delta = 3.9$ as a starting function.

function, while the spin correlations for the LA-PHF wave function remains still almost zero. Moreover in the range $\Delta \leq 4$ we obtain the state where the squared spin is larger than 0.75(this value corresponds to $1\mu_{\rm B}$ as the magnetic moment), when we use the wave function corresponding to a larger Δ as a starting function. The variational parameters η_1 and η_2 are about 0.2 and -0.8 respectively for the all charge transfer energy range investigated in the calculation. The Hund rule coupling correlation $\vec{s}_u \cdot \vec{s}_v$ is still nearly zero($\leq 10^{-1}$) even after including the correlation.

5. Concluding Remarks

The above-mentioned result for the linear chain model that Ni atoms in the metallic site keep a fairly large magnetic moment near the metal-insulator transition seems to be quite consistent with the photoemission experiment on NiS. However, the fact that the size of the model is too small casts some doubt on the validity. Also the calculation results in various spurious transitions arising from the finite size. The increase of the antiferromagnetic spin correlation between a Ni atom and neighboring ligand atoms in the intermediate phase (M) suggests that the spin of a Ni atoms locally compensated to some extent by the spins on neighboring ligand atoms; the decrease of the spin correlation between Ni atoms and the disappearance of the Hund rule coupling are consistent with this picture. Compared with the phase (I) which certainly corresponds to the insulating phase, the number of holes on the ligand atoms increases in the phase (M). We can imagine that the phase (M) resembles the concentrated Kondo phase.

From the preliminary Local Ansatz calculation reported above we conclude that indeed the metallic state in the low charge transfer energy region($0 \leq \Delta \leq 4$) will be modified by the electron correlation to result in a local moment larger than $1\mu_B$ for a Ni atom, and a small antiferromagnetic correlation with neighboring Ni moments; the calculation shows also that the Hund rule spin correlation within a Ni atom remains small in the metallic state. We cannot pin-point the value of the charge transfer energy where the transition into the insulating phase will occur because of the lack of accuracy of the calculation of the total energy. Nevertheless we may conclude that electrons in the metallic phase near the transition is highly correlated to make the ordinary band structure calculation inapplicable.

It is highly desirable to carry out the photoelectron spectroscopy experiment with various wave lengths of light or the resonant photoemmision spectroscopy experiment to check the change of the electronic structure through the transition in more detail. Our calculation suggests that peak heights will be dependent on the wave length through the difference of the energy dependence of the transition matrix elements between Ni and ligand atom, while peak positions do not change appreciably. Also the measurement of the magnetic moment of Ni atom in the metallic phase by neutron will be useful to understand what is going on there.

Our linear model calculation simulating a Cu compound shows a gradual change of the ground state when the charge transfer energy is varied. The abrupt change of the ground state in the Ni case seems to originate in the local stability of the insulating antiferromagnetic phase for which the Hund rule spin correlation among different d orbitals might be responsible. We shall elucidate this point in future analysis.

Acknowledgment

We would like to express our sincere thanks to Dr.T.Nishino, Dr.M.Kikuchi and other members of our laboratory for their valuable discussions. The present research was supported by Grants-in-Aid for Scientific Researches, in particular Grant-in Aid No.01540275 of General Scientific Research and Grant-in-Aid of Scientific Research on Priority Areas 031 from Ministry of Education, Science and Culture.

References

[1] R. F. Koehler Jr. and R. L. White: J. Appl. Phys. **44** (1973) 1682.

[2] A. Fujimori, M. Matoba, S. Anzai, K. Terakura, M. Taniguchi, S. Ogawa and S. Suga: J. Mag. and Mag. Mat. **70** (1987) 67. A.Fujimori, H.Namatame, M.Matoba and S.Anzai: Phys. Rev. **B42** (1990) 620.

[3] A. Fujimori, K. Terakura, M. Taniguchi, S. Ogawa, S. Suga, M. Matoba and S. Anzai: Phys. Rev. **B37** (1988) 3109.

[4] A. Fujimori, F. Minami and S. Sugano: Phys. Rev. **B29** (1984) 5225.

[5] G. A. Sawatzky and J. W. Allen: Phys. Rev. Lett. **53** (1984) 2239.

[6] M. Takahashi and J. Kanamori: J. Phys. Soc. Jpn. **60** (1991) 3154.

[7] P. Fulde: *Electron Correlations in Molecules and Solids*, Springer Ser. Solid-State Sci., Vol.100 (Springer-Verlag 1991)

[8] C.Lanczos: J. Res. Nat. Bur. Standards **45** (1950) 255.

[9] H. Yokoyama and H. Shiba: J. Phys. Soc. Jpn. **56** (1987) 1490.

[10] R. M. Fye and J. E. Hirsch: Phys. Rev. **B40** (1989) 4780

[11] A. M. Olés: J. Phys. **C15** (1982) 2745.

[12] J. Zaamen, G. A. Sawatzky and J. W. Allen: J. Mag. and Mag. Mat. **54-57** (1986) 607, Phys. Rev. Lett. **55** (1985) 418.

[13] K. Yosida and S. Inagaki: J. Phys. Soc. Jpn. **50** (1981) 3268.

[14] J. van Elp, H. Eskes, P. Kulper, and G. A. Sawatzky: Phys. Rev. **B45** (1992) 1612.

[15] L. F. Mattheiss: Phys. Rev. **B5** (1972) 290.

Index of Contributors

Springer Series in Solid-State Sciences

Editors: M. Cardona P. Fulde K. von Klitzing H.-J. Queisser

1 **Principles of Magnetic Resonance** 3rd Edition By C. P. Slichter
2 **Introduction to Solid-State Theory** 2nd Printing By O. Madelung
3 **Dynamical Scattering of X-Rays in Crystals** By Z. G. Pinsker
4 **Inelastic Electron Tunneling Spectroscopy** Editor: T. Wolfram
5 **Fundamentals of Crystal Growth I** Macroscopic Equilibrium and Transport Concepts. 2nd Printing By F. Rosenberger
6 **Magnetic Flux Structures in Superconductors** By R. P. Huebener
7 **Green's Functions in Quantum Physics** 2nd Edition By E. N. Economou
8 **Solitons and Condensed Matter Physics** 2nd Printing Editors: A. R. Bishop and T. Schneider
9 **Photoferroelectrics** By V. M. Fridkin
10 **Phonon Dispersion Relations in Insulators** By H. Bilz and W. Kress
11 **Electron Transport in Compound Semiconductors** By B. R. Nag
12 **The Physics of Elementary Excitations** By S. Nakajima, Y. Toyozawa, and R. Abe
13 **The Physics of Selenium and Tellurium** Editors: E. Gerlach and P. Grosse
14 **Magnetic Bubble Technology** 2nd Edition By A. H. Eschenfelder
15 **Modern Crystallography I** Symmetry of Crystals. Methods of Structural Crystallography By B. K. Vainshtein
16 **Organic Molecular Crystals** Their Electronic States. By E. A. Silinsh
17 **The Theory of Magnetism I** Statics and Dynamics. 2nd Printing By D. C. Mattis
18 **Relaxation of Elementary Excitations** Editors: R. Kubo and E. Hanamura
19 **Solitons.** Mathematical Methods for Physicists. 2nd Printing By G. Eilenberger
20 **Theory of Nonlinear Lattices** 2nd Edition By M. Toda
21 **Modern Crystallography II** Structure of Crystals By B. K. Vainshtein, V. M. Fridkin, and V. L. Indenbom
22 **Point Defects in Semiconductors I** Theoretical Aspects By M. Lannoo and J. Bourgoin
23 **Physics in One Dimension** Editors: J. Bernasconi, T. Schneider
24 **Physics in High Magnetic Fields** Editors: S. Chikazumi and N. Miura
25 **Fundamental Physics of Amorphous Semiconductors** Editor: F. Yonezawa
26 **Elastic Media with Microstructure I** One-Dimensional Models. By I. A. Kunin
27 **Superconductivity of Transition Metals** Their Alloys and Compounds By S. V. Vonsovsky, Yu. A. Izyumov, and E. Z. Kurmaev
28 **The Structure and Properties of Matter** Editor: T. Matsubara
29 **Electron Correlation and Magnetism in Narrow-Band Systems** Editor: T. Moriya
30 **Statistical Physics I** Equilibrium Statistical Mechanics 2nd Edition By M. Toda, R. Kubo, N. Saito
31 **Statistical Physics II** Nonequilibrium Statistical Mechanics 2nd Edition By R. Kubo, M. Toda, N. Hashitsume
32 **Quantum Theory of Magnetism** 2nd Edition By R. M. White
33 **Mixed Crystals** By A. I. Kitaigorodsky
34 **Phonons: Theory and Experiments I** Lattice Dynamics and Models of Interatomic Forces. By P. Brüesch
35 **Point Defects in Semiconductors II** Experimental Aspects By J. Bourgoin and M. Lannoo
36 **Modern Crystallography III** Crystal Growth 2nd Edition By A. A. Chernov
37 **Modern Crystallography IV** Physical Properties of Crystals Editor: L. A. Shuvalov
38 **Physics of Intercalation Compounds** Editors: L. Pietronero and E. Tosatti
39 **Anderson Localization** Editors: Y. Nagaoka and H. Fukuyama
40 **Semiconductor Physics** An Introduction 5th Edition By K. Seeger
41 **The LMTO Method** Muffin-Tin Orbitals and Electronic Structure By H. L. Skriver
42 **Crystal Optics with Spatial Dispersion, and Excitons** 2nd Edition By V. M. Agranovich and V. L. Ginzburg
43 **Structural Analysis of Point Defects in Solids** An Introduction to Multiple Magnetic Resonance Spectroscopy By J.-M. Spaeth, J. R. Niklas, and R. H. Bartram
44 **Elastic Media with Microstructure II** Three-Dimensional Models By I. A. Kunin
45 **Electronic Properties of Doped Semiconductors** By B. I. Shklovskii and A. L. Efros
46 **Topological Disorder in Condensed Matter** Editors: F. Yonezawa and T. Ninomiya
47 **Statics and Dynamics of Nonlinear Systems** Editors: G. Benedek, H. Bilz, and R. Zeyher
48 **Magnetic Phase Transitions** Editors: M. Ausloos and R. J. Elliott
49 **Organic Molecular Aggregates** Electronic Excitation and Interaction Processes Editors: P. Reineker, H. Haken, and H. C. Wolf

Springer Series in Solid-State Sciences

Editors: M. Cardona P. Fulde K. von Klitzing H.-J. Queisser

Springer-Verlag and the Environment

We at Springer-Verlag firmly believe that an international science publisher has a special obligation to the environment, and our corporate policies consistently reflect this conviction.

We also expect our business partners – paper mills, printers, packaging manufacturers, etc. – to commit themselves to using environmentally friendly materials and production processes.

The paper in this book is made from low- or no-chlorine pulp and is acid free, in conformance with international standards for paper permanency.

MIX
Papier aus verantwortungsvollen Quellen
Paper from responsible sources
FSC® C105338

If you have any concerns about our products,
you can contact us on
ProductSafety@springernature.com

In case Publisher is established outside the EU,
the EU authorized representative is:
Springer Nature Customer Service Center GmbH
Europaplatz 3, 69115 Heidelberg, Germany

Printed by Libri Plureos GmbH
in Hamburg, Germany